CONTENTS

EARTH'S
RESTLESS SURFACE

Deirdre Janson-Smith
with Gordon Cressey and Andrew Fleet

Publi   London

First published by the Natural History Museum, Cromwell Road, London SW7 5BD
© Natural History Museum, London, 2008

ISBN 13 978 0 565 09236 8

Designed by Mercer Design
Reproduction by Colourscan Overseas Co Pte Ltd
Printing by Craft Print International Ltd

Front cover image: Mount Roraima, Venezuela
Back cover images: Hurricane Katrina, an uprooted tree sits atop a car and house in Ward 9, New Orleans, USA; The Mahakam Delta in East Kalimantan; Tre Cime di Lavaredo, Italy; Sahara Desert, North Africa; Salt formations extend to the horizon at Badwater in Death Valley, USA; The Twelve Apostles along the Great Ocean Road, Australia; Colorado River, Grand Canyon, USA.

Acknowledgements: The authors would like to thank Sharon Fieldman, Barney Smith and Dr Sue Rigby for their help.

PREFACE

THIS BOOK IS ABOUT CHANGE AT THE EARTH'S SURFACE. It explores the processes and products of change. It explains how agents of erosion sculpt the landscape and how their action in turn leads to the formation of other rocks. It shows how rocks can be read as a record of past processes, and how we can extend our interpretation back through time to recreate whole landscapes.

The surface of our planet is in flux, constantly being remodelled by powerful natural forces. Violent storms and hurricanes, earthquakes and volcanoes are not occasional upsets, but part and parcel of the Earth's basic nature. We need to understand our planet as a dynamic, interacting system that responds to change in highly complex and as yet poorly understood ways.

- How important is the Earth's atmosphere?
- How do solid rocks wear away, dissolve and become part of new rocks elsewhere?
- How are mountains levelled?
- What are the driving forces behind the erosive power of water, wind and ice?
- How are rocks and minerals recycled at the Earth's surface and into its interior?
- And what is the human impact on natural cycles and climate change?

This book gives some flavour of what scientific research is revealing about how change happens on local and global scales, and how the great interlocking cycles react and interact to achieve a dynamic equilibrium. This story of natural processes also includes what can happen if human actions disturb the dynamic rhythms of change and exchange.

OPPOSITE Devastation to villages in the Irrawaddy delta following the May 2008 Cyclone Nargis which hit Burma (Myanmar).

INTRODUCTION

O N 26 DECEMBER 2004, AN EARTHQUAKE off the coast of Indonesia triggered a massive tsunami that left nearly 230,000 people dead or missing, and another 2 million homeless. This natural catastrophe was a shocking reminder of the power of natural forces. Tsunami with this destructive force are rare on the human timescale, but in some regions destruction by events such as hurricanes and floods are commonplace. Even outside these areas, everywhere the power of wind and water, ice and moving rocks is gradually transforming the surface of the Earth.

The Earth's surface has seen constant change throughout its history of 4567 million years. Violent sudden activity in the form of earthquakes and tsunamis, landslides and floods punctuates change. While infinitesimally slow and small actions, over geological stretches of time, mould the landscape.

OPPOSITE The December 2004 Indian Ocean tsunami was caused by an underwater earthquake just off the northwest Indonesian coast. The tsunami caused devastation along the coasts of Thailand, Indonesia and stretching to Sri Lanka and the east coast of India.

LEFT An uprooted tree sits atop a car and house in Ward 9, New Orleans following Hurricane Katrina in August 2005.

Two major driving mechanisms together bring about this change: the sun's radiant energy reaching the Earth's surface, and the heat produced by radioactive decay in the Earth's interior. This book deals primarily with the first of these – the sun as an engine of change – and investigates the agents that remove and redistribute surface material, which are ultimately driven by the input of solar energy. Gravity also plays a key role; potential energy arising from the gravitational attraction of the Earth, fuels the down-slope flow of water, ice, rock and soil. Under these two forces the agents of change attack and erode the Earth's surface, to reduce its elevation and relief.

But the action of surface forces is not sufficient to explain the shaping of the landscape over time. It is also necessary to take into account the internal engine of change which acts in opposition to them, creating and building up the surface. And we need to understand how these two opposing forces interact to shape the Earth's surface over time.

BELOW The Earth's outer layer (lithosphere) consists of 8 major tectonic plates. Their formation, movement apart and collision generates unimaginable forces, opening and closing oceans and building mountains.

THE INTERNAL ENGINE

The 1960s and 1970s saw a dramatic advance in our understanding of the Earth as the concept of plate tectonics was developed and refined. We now know that the outermost solid layer of the Earth, the lithosphere, is not continuous but is

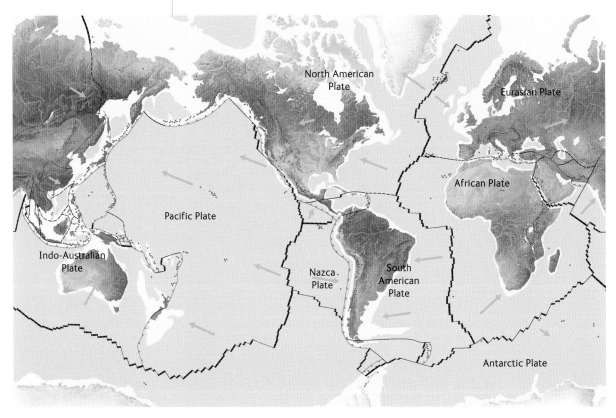

North American Plate

Eurasian Plate

African Plate

Pacific Plate

Indo-Australian Plate

Nazca Plate

South American Plate

Antarctic Plate

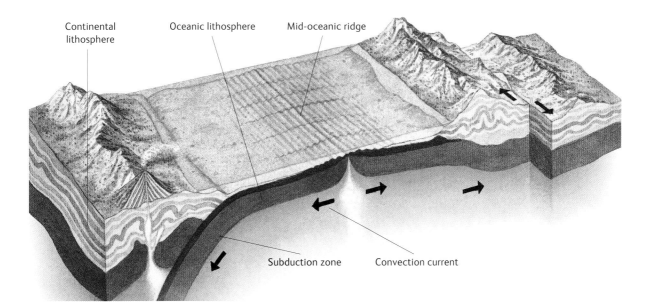

Continental lithosphere

Oceanic lithosphere

Mid-oceanic ridge

Subduction zone

Convection current

ABOVE New oceanic lithosphere is continually being generated along mid-oceanic ridges; at some continental margins it sinks into the Earth's interior at subduction zones.

constructed of giant moving plates. The lithosphere is made up of the Earth's crust and outermost part of its mantle. Where the crust is relatively dense it forms the basins that are occupied by the oceans. The continents occur where the crust is less dense. (The continental crust may be as much as 70 km (43½ miles) thick below the highest mountains, while oceanic crust is rarely more than 6 km (3¾ miles) thick.) The plates are in constant motion, driven by convection currents in the hot, plastic region of mantle (the asthenosphere) that immediately underlies them. Where the plates collide, one is drawn down under the other, recycling material into the interior, causing rocks to melt and rise as volcanic magma to the surface. Where the plates pull apart, rising magma solidifies to create new sea floor at giant mid-oceanic ridges.

Plate movements are responsible for the massive uplift and buckling of the land surface over huge periods of time, as well as for sporadic volcanic eruptions and earthquakes. It is this internal engine of the Earth that creates the land relief which will in turn be acted on by atmospheric forces at the surface.

STILL MOON

The moon is our nearest neighbour, so near that we can see some of its surface features with the naked eye. One current theory of its origins suggests that a Mars-sized planetary body hit the Earth with a glancing blow some 4500 million years ago, flinging out the matter to form the moon. The moon's gravity is only 1/6 of that on Earth, and so it lacks the gravitational pull to hold onto an atmosphere. It also lacks water and while 17% of its surface is covered by ancient lava flows, volcanic activity probably ceased about 3.9–3.1 billion years ago. So, to all intents and purposes, the surface of the moon is dead. All the forces which mould the Earth's surface are missing.

ABOVE The moon's surface is heavily pockmarked by the impact of meteorites. Nothing else disturbs it.

WHAT DO YOU MEAN, 'SURFACE'?

In this book, the surface of the Earth is taken to mean not just the solid crust, but also the envelope of gases that protect it, the waters that flow across and through it, and the living organisms that live on and in it. These four systems interact with one another in complex ways that we are only just beginning to understand. They exchange matter and energy in continuous cycles of change over time. They are in a state of dynamic equilibrium that shifts its balance over geological time.

THE SOLID CRUST

The solid surface of the Earth comprises the soil and the rocks that lie below the land and beneath the ocean water. Plate tectonics generates new surface rocks from the interior, and processes at the surface reshape these, fragmenting, eroding, transporting and depositing them again to form new rocks.

RIGHT The solid crust, atmosphere and hydrosphere which together with all the living organisms (the biopshere) make up the 'surface' of the Earth.

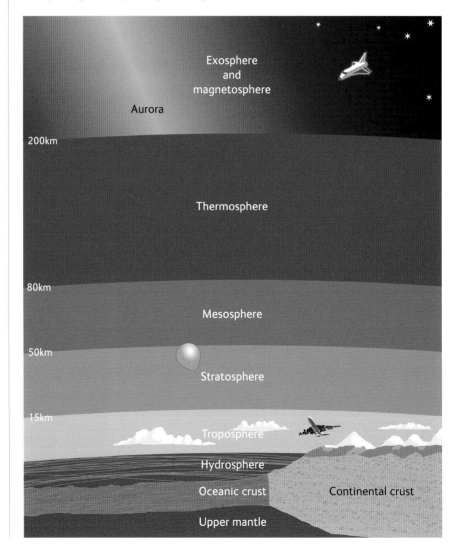

Exosphere and magnetosphere

Aurora

200km

Thermosphere

80km

Mesosphere

50km

Stratosphere

15km

Troposphere

Hydrosphere

Oceanic crust

Continental crust

Upper mantle

LEFT Solid crust – surface processes destroy rocks and transport and deposit the fragments formed giving rise to new rocks.

THE ATMOSPHERE

The atmosphere is a thin blanket of gases, primarily nitrogen and oxygen, held to the Earth by gravitational attraction. Without it, our world would be as lifeless as the moon. The atmosphere keeps us warm, shields us from the sun's harmful rays and cycles warmth, water and other chemicals. It has a complex series of layers stretching some 1000 km (621 miles) out into space, although at this distance very few molecules can be detected. A layer called the troposphere contains 99% of the atmospheric gases, concentrated in the first 15 km (9 miles). Virtually all water vapour and clouds exist in this layer, and almost all weather occurs here. (Weather is essentially the atmosphere in turbulent motion.)

By reflecting or absorbing and then transporting the sun's energy, the atmosphere helps the planet balance its energy budget and maintain a global average air temperature at the surface of about 14°C (57°F). The atmosphere allows sunlight through, while gases in the atmosphere such as carbon dioxide and water vapour trap heat radiating back from the surface (the greenhouse effect, see p.104). Without this barrier, the heat loss would be so rapid that the Earth's surface would cool drastically when not directly warmed by the sun (the moon's 'dark side' is minus 120°C (248°F)). The Earth's atmosphere has evolved with the evolution of life.

ABOVE Atmosphere – the thin veil of gases that both protects and interacts with the surface.

THE HYDROSPHERE

About 70% of the planet is veiled in water, which makes it unique in our solar system. Water circulates between air, land and oceans. (This pattern of movement, called the hydrologic or water cycle is shown on p.71) Oceans hold 97.3% of the planet's 1.3 billion cubic kilometres of water. A further 2.1% is in ice sheets and glaciers – enough for sea levels worldwide to rise an additional 65 m (213 ft) if this ice were to melt. Groundwater accounts for about 0.6% while 0.01% is in freshwater streams, lakes and rivers. A tiny but critical fraction (0.001%) is in the atmosphere as water vapour. This still amounts to 4000 million tonnes of water, raining on the land and sea every year. Water currents moderate weather and climate by transporting heat from the equator towards the Poles. And moving water erodes, transports and deposits material to shape the solid surface.

ABOVE Biosphere – life inhabits all three other systems of the Earth's surface even occurring down to 1 km (1/2 mile) or so in the solid crust.

THE BIOSPHERE

The biosphere is the 'layer' at the surface containing all life, spread between the other layers of air, water and earth with microbes, down to a depth of a kilometre or more in the solid crust. Living things play a vital part in the cycling of materials and energy through the surface systems. They also create the conditions for their own survival; it was the development of oxygen-releasing organisms that led to the build-up of free oxygen in the atmosphere, on which most life-forms now depend, and which is maintained by living things.

INTERACTING SYSTEM

While each of these four systems can be described separately, it is essential to regard the surface as a single, interacting system. The behaviour of the air and water and of living organisms determines surface processes: flowing water, winds and ocean currents create patterns of erosion, transport and deposition to shape the surface. And in turn, the form of the surface influences these processes: the position of the continents; the relative position of land and sea; and, the height and form of mountains all affect the nature of the action of wind and water currents.

DRIVEN BY THE SUN

At the centre of our solar system, the sun pours out vast quantities of energy into space. We receive only a tiny fraction of this energy on Earth, perhaps only one billionth, but it is sufficient to affect the form and nature of our planet radically. Solar radiation provides the energy for biological activity, the evaporation of water and drives the global circulation of the atmosphere and oceans.

BELOW Energy streaming from the sun warms the Earth and drives its surface processes.

WHY IS IT HOTTER AT THE EQUATOR?

Imagine a torch shining on a flat surface. If the light is held directly over the surface and the beam shines straight down, it illuminates a small area brightly. This is the situation at the equator.

If the torch is held at an angle, a larger area is illuminated, but the intensity of the light is less. One unit of light is spread over a larger area. This is the situation at high latitudes of the globe.

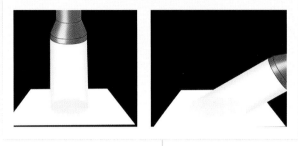

The sun shines unevenly on the Earth's surface and this simple fact is critical to all that follows. When the sun shines on the spherical Earth, the equator receives the most concentrated energy, and is generally warm throughout the year. In winter, regions at high latitudes are at such an angle to the sun that it barely rises, if at all, above the horizon there. Polar regions are cold and ice-bound even in summer. The sun's radiation falls on the Earth's surface unevenly, setting in motion a restless movement of air and water to distribute this energy across the planet from the warm equator to cold Poles, in great convection currents. If air in one region of the globe is heated above the temperature of surrounding air, the warm air becomes less dense and rises. As the warm air rises, cooler, denser air in another portion of the atmosphere sinks. Air flows along the surface to complete the cycle. The steady winds that blow across the tropical oceans are ultimately caused by these churning convection currents.

It might be expected that there would be two giant convection currents, one either side of the equator, with hot air rising at the equator and cold air sinking at the Poles – a basic pattern of circulation proposed by the British scientist George Hadley in 1735. In fact, there are six circulatory cells of air motion. Either side of the equator, air sinks at about 30° latitude, then splits in two and flows north and south from this latitude to create typical patterns of prevailing winds. The winds change direction with the seasons, as different patterns of flow are set up. Their flows determine broad climatic zones worldwide. Hadley cells are created because the winds are deflected by the spinning of the Earth. Winds flowing from equator to Pole are deflected in the northern hemisphere to the right, and in the southern hemisphere to the left. Winds flowing from north and south towards the equator are deflected in the opposite directions. This deflection is called the Coriolis effect.

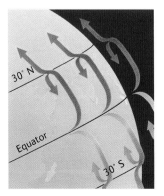

ABOVE Simple convection current – as water is heated, it becomes less dense and rises; cooler, more dense water sinks.

BELOW Convection currents in the atmosphere transfer heat away from the equator.

CLIMATE

Weather is essentially the result of the churning of the atmosphere. And the average type of weather over a period of time is known as climate. Because of the global flow of air currents either side of the equator (described above), areas on the same latitude generally experience broadly similar climatic conditions. We can group the Earth's major climate zones into just a few major 'belts' classified by temperature and precipitation: polar regions with cold, and dry conditions; temperate regions, with cool and wet conditions; desert regions, with hot and dry conditions; and tropical regions, with hot and wet conditions.

However, so many other factors affect regional and local climate that it cannot be predicted by latitude alone. For example, the oceans' ability to retain heat more effectively than air means that coastal climates are generally, although not always, warmer and moister. Mountain ranges also alter the movement of air and generate local climate zones, and cold, dry conditions often result in a glacial environment, where ice is the major agent of erosion and transport. In desert environments, under hot, dry conditions, wind is the major agent of erosion and transport. And where temperatures and rainfall are both high, in tropical regions, the chemical breakdown of rocks is speeded up.

CURRENTS OF WATER

Ocean currents are driven by the winds of the atmosphere, because when wind blows across water in a constant direction, it drags the water with it. The major global currents move in circular paths called gyres. They too are deflected by the Coriolis effect and rotate clockwise in the northern hemisphere, and anti-clockwise in the south. Their pattern of flow is further determined by the arrangement of the continental landmasses.

Like the winds, the oceans transport large quantities of heat northwards and southwards, away from the equator. For example, the Gulf Stream of the Atlantic, which becomes the North Atlantic Drift before it reaches Europe, carries one million cubic metres of warm water every second, warming coastal areas. This is why Glasgow in Scotland is relatively mild in comparison with Churchill in Canada,

BELOW Broad climate zones are created by airflow in the atmosphere either side of the equator.

North Pole

Polar

Temperate

Desert

Tropical

Equator

THE SEASONS

The Earth rotates around the sun once a year. Because it is tilted on its axis, northern and southern hemispheres receive uneven amounts of sunlight at different times of the year. The North Pole tilts towards the sun in the northern hemisphere's summer, and away from it in winter. This means that in summer the sun is relatively high in the sky; the sunlight is more concentrated and is there for a longer part of the day. (When the North Pole is tilted towards the sun it receives 24 hours of daylight: the 'midnight sun'). This uneven, changing energy input is the cause of the seasons with their changing patterns of winds and currents.

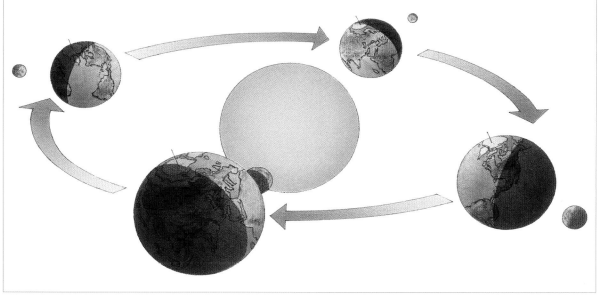

RIGHT Like giant rivers in the sea, surface currents flow around the oceans. And like the prevailing winds, ocean currents are deflected by the spinning of the Earth.

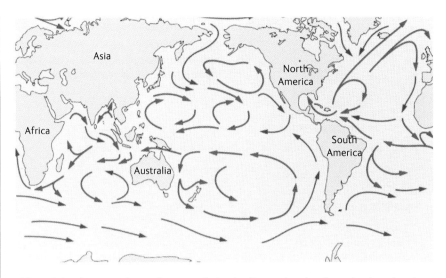

BELOW Coriolis effect – the rotation of the Earth on its axis deflects air and sea currents from their straight paths.

although both are at about the same latitude. (It used to be thought that the deep sea was motionless, unaffected by the surface winds. Now we know that water flows here too. Its movement is driven by turbulence caused by winds and tides and differences in water density.)

As well as redistributing the sun's energy, wind and water also have the ability to erode, transport, and deposit sediments throughout the globe, and are thus the major players in the story of the Earth's restless surface.

DRIVEN BY THE MOON

The oceans also move in tides, a rhythm of movement in response to the gravitational attraction of the moon, and to a lesser extent of the sun. (Although the moon is much smaller than the sun, it is so much closer to the Earth that its gravitational influence is stronger.) At any given time, one point of the Earth is directly under the moon and that part of the ocean is attracted with the strongest force. This attraction causes the ocean to bulge outwards towards the moon, resulting in a high tide – as the Earth rotates, the bulge always faces the moon. When both sun and moon are directly in line with the Earth, their gravitational fields reinforce each other to create a strong tidal bulge, called a spring tide. When they are 90° out of alignment, each partially offsets the effect of the other and the smallest tidal differences are observed, called neap tides.

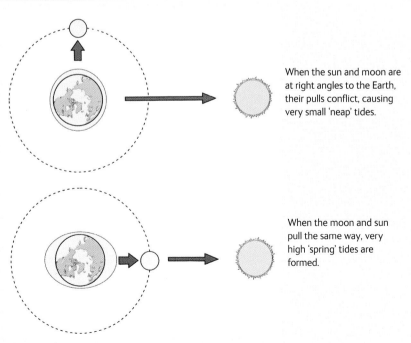

When the sun and moon are at right angles to the Earth, their pulls conflict, causing very small 'neap' tides.

When the moon and sun pull the same way, very high 'spring' tides are formed.

TIME AND CHANGE

The concept of time is central to geological thought. The processes that shape the surface of the Earth operate over vast expanses of time – millions or even billions of years. If we discount, for the moment, catastrophic events such as earthquakes and landslides, the Earth's surface seems relatively stable over the timescales we can measure. Historical records do show the slow diversion of a river's course, and the silting up of estuaries, for example, but usually we would expect images taken a hundred years ago to show a landscape essentially the same as today's.

But imagine time speeding up, so that a million years pass in a minute. In this timeframe, we would soon lose our concept of 'solid' Earth, as we watch the restless surface change out of all recognition. There is virtually no place on the Earth's surface that is not moving either vertically or horizontally. The Himalayas began to rise about 50 million years ago, and are still rising today. Scandinavia, too, is rising at the rate of about 1 cm (1/2 inch) per year, rebounding from being weighed down by a vast ice sheet 2–3 km (1-2 miles) deep which covered it 40,000 years ago. At the same time erosion is also working slowly and inevitably to lower the land. North America's land surface is eroding away at an average of 0.03 mm (1/10 inch) per year and the Niagara Falls are being cut back by the force of the waterfall at a rate of over a metre a year.

BELOW The path a river cuts over thousands of years can change in a few hours. Five sketches showing how the path of a river changes over time.

Meander accentuated over thousands of years

Oxbow lake cut off after one hour of heavy flood.

LEFT The Grand Canyon, USA cut by the Colorado River over the last 20 million years.

Some processes work so slowly that, to our eyes, nothing has changed. The hammering of rain, the molecule-by-molecule dissolution of a rock, the slow scouring of a glacier – all act slowly to wear down the land. Measured on a grand scale, a mountain chain may take ten million years to build, and ten million years to erode.

Many events are not continuous but periodic. Spring floods will wash down large quantities of fresh sediment to a lake, then the supply dwindles for the rest of the year. And sometimes change comes with terrifying speed. A sudden flash flood can accomplish in an hour what slow erosion has taken tens of thousands of years to achieve. Over geological time, each of these are important. Each can alter whole landscapes beyond recognition.

ABOVE James Hutton was among the first to suggest that natural agents shaped the Earth's surface.

CHANGING CONCEPTS

It is only in the last two hundred years or so that we have come to terms with the vastness of geological time. At the end of the 18th century, common belief in the West still held that the Earth had been shaped by catastrophic events as described in the Bible, and most people thought that the Earth was only 6000 years old. But geologists studying rock formations in the field were piecing together a very different story, with a very different timescale.

James Hutton, an Edinburgh geologist of the late 18th century, was among the first to argue that natural agents such as flowing water, acting slowly over thousands or even millions of years, shaped the Earth's surface. Furthermore, he proposed that

the surface went through a great cycle of change with 'no vestige of a beginning, no prospect of an end'. This cycle was, he suggested, driven by the erosive power of rivers that wore down mountains and washed down sediment to the sea to form new rocks. Then a great upheaval in the surface thrust those same rocks up once more, and created new (igneous) rocks, to continue the cycle.

The argument that neither cataclysmic forces nor divine intervention were necessary to explain the structures of the surface was developed further by Charles Lyell, whose 1830 publication, *Principles of Geology*, still guides geological thought. In this, Lyell formulates the principle of uniformitarianism. Simply put, it states that in geology we should assume that natural laws are constant in time and space, and that the processes we see in action today are sufficient to explain the past – that 'the present is the key to the past'. (Lyell's theory is more complex and subtle than this, and it is worth reading Stephen Jay Gould's *Time's Arrow, Time's Cycle* for an exploration of both his and Hutton's thoughts.) Uniformitarianism is still a key idea in geological thought today. As with all explanations in science, it has undergone modification as we have learned more about the Earth's processes. Hutton and Lyell certainly had no knowledge of fundamental processes of plate tectonics, nor of climate change, for example. And geologists today are beginning to recognise the importance of change at many different rates. The effects of hurricanes, catastrophic landslides and slow creeping of land on the form of the surface can be measured during a human life. The relative impact, though, of a major tsunami or volcanic eruption occurring once every hundred thousand or million years is more difficult to assess.

SCALE AND CHANGE

Geology is concerned with processes that act at many different scales, from the structure of minerals to the global budget of whole systems. The question of scale is intimately bound up with the concept of time. Small events at a micro-scale can be studied within a limited timeframe; expand the scale, and major changes can only be measured across geological time.

At different scales, causality becomes more or less complex. In a mountain stream, the wearing down of a boulder and its ultimate fate can be explained by a few factors. But the formation of a valley has many and complex causes. The formation of a mountain range has so many that its pattern of erosional development may essentially be unpredictable.

Studied on this wider scale, small variations in the nature and rate of change become insignificant – just as a distant view of a hillside describes its shape but fails to reveal the roughness of its contours. We may observe radical differences in the behaviour of a river from spring flood to high summer drought, and the kind of change it brings about. But take a step back and analyse the river system averaged over thousands of years and these variations will be less significant in our description of how it has changed. The two concepts of scale and time are closely connected.

RIGHT If a rock is coarse-grained, individual mineral grains can be seen in a 'hand specimen' and studied with a hand lens.

OPPOSITE On a wider scale, we can study the nature of distinctive landscape and their evolution over geological time, like these mountains and valleys of Yosemite Falls and the Merced River, Sierra Nevada, California, USA.

BELOW Slices of rock ground to 30 microns thickness are thin enough to transmit light so that the minerals which they contain and their structure can be studied using a microscope, even if the rock consists of grains too fine to be seen by the naked eye.

Modern geological research is increasingly concerned with understanding the complex dynamics of the whole-Earth system and how the solid crust, atmosphere, hydrosphere and biosphere interact. This change in perspective has been made possible because of three key factors. Firstly, since the 1960s, plate tectonics has had a major impact on our understanding of the long-term dynamics that shape the Earth's surface. Secondly, since the 1970s, we have begun to explore the whole surface of the Earth, through exploration of the deep-ocean floor, that covers two thirds of the surface, and through satellite imaging and data gathering. Thirdly, developments in research techniques and in computer modelling are increasingly allowing data to be gathered, integrated and analysed from the atomic to the global scale.

ROCKS AND MINERALS

At micro-scale, rocks can be seen to be composed of smaller units – mineral grains. A mineral is a naturally occurring chemical compound of particular composition and with a specific 'architectural' arrangement of its constituent atoms. Minerals are made from elements, basic chemicals such as silicon and oxygen (which together make up 75% of the Earth's crust). Some minerals have complex chemical compositions, others are simple. Quartz, for example, is made of silicon and oxygen. Graphite and diamonds contain only carbon, but each has a

different atomic architecture. To date, about 4000 different minerals have been discovered but most are very rare. Only 30 or so are common at the Earth's surface.

Minerals are the building blocks of rocks. Most rocks contain several minerals, often six or more. Some contain just one. In coarse-grained rocks, the component minerals are easy to see. Granite, for example, a common rock exposed at the Earth's surface, is made essentially of three minerals: quartz, feldspar and mica. Granite is formed by slow cooling in the Earth's interior. It is this slow cooling rate that allows coarse-grained crystals, visible to the naked eye to form.

FAR LEFT This unprepossessing white mineral, sent to the Natural History Museum, London in 2007 turned out to be a new mineral that was named jaderite. Research showed that it had the same chemical composition as that given to the fictional mineral kryptonite in the Superman adventures.

LEFT In 'mineral' water, there are simply no minerals at all. The water contains a mix of compounds in solution (dissociated into ions).

WEATHERING

THIS CHAPTER IS ABOUT BEGINNING THE PROCESS of transformation at the Earth's surface. The surface may seem 'solid as a rock', but over thousands of years even the hardest rocks will be worn away. They are physically and chemically attacked at the surface in a process called weathering, just as an iron nail outdoors becomes rusty and fragile and finally turns to dust.

There are two major categories of weathering, although in reality they interact considerably. Chemical weathering is the chemical breakdown, sometimes total dissolution, of rocks. Physical weathering is the physical break-up of a rock into smaller fragments. In both cases, weathering is essentially a process of readjustment of minerals which formed under very different environmental conditions to those prevailing at the Earth's surface. For minerals that were formed deep inside the Earth, the surface is a very different place. Some minerals remain very stable on exposure, while others are transformed, or dissolve totally away.

OPPOSITE The dramatic tower karst scenery around the Li River, Guilin, southwest China, forms steep-sided mountains in metamorphic limestone. The rock is heavily chemically weathered and eroded under humid, subtropical conditions.

CHEMICAL ATTACK

At the Earth's surface, unlike the interior, there is abundant water and free oxygen, both of which can destabilise minerals. In an oxygen-rich atmosphere, rocks with a high iron content will 'rust', oxygen reacts with the iron to form iron oxides (rust), which often remain behind as a red or brown stain in rocks and soils as other elements are removed in solution. Rocks that contain other minerals will be similarly converted to different coloured rusts, for example, copper rust – 'verdigris' – is green.

Given enough time, pure water can dissolve any rock; its ability to do so will depend on the nature of the minerals that make up the rock. But the ability of water to attack certain types of rocks increases immensely if it is slightly acidic. Surface water contains a tiny amount of dissolved carbon dioxide which combines with it to form carbonic acid.

The breakdown of granite is largely due to one chemical reaction – between feldspar and acidified water. Feldspar is a key mineral in the Earth's crust, composed of potassium, sodium, silicon, oxygen and aluminium. Water decomposes feldspar to kaolinite (the basis of China clay or kaolin) by removing the potassium and sodium and leaving behind an aluminium silicate. Instead of the crystalline feldspar, weathered granite contains loosely adhering kaolinite, and its structure falls apart, releasing the other mineral grains, mica and quartz. (Quartz is very resistant to weathering, both chemical and physical, which is why many sandstones are almost entirely quartz.)

ABOVE Many stone buildings are being eroded by acid attack from industrial pollution.

RIGHT Weathered granite is crumbly and discoloured, altered by chemical attack.

LEFT Lichens growing over a rock surface secrete acids that slowly dissolve away the rock.

ABOVE Rock-boring molluscs abrade and dissolve rocks with their acid secretions.

Limestone is a tough, fairly resistant rock, forming major landscapes. However, it will dissolve away totally on contact with acid rainwater. This chemical reaction turns the calcium carbonate of limestone into soluble calcium bicarbonate and releases carbon dioxide gas. In turn, carbon dioxide is removed from the atmosphere by dissolving in rainwater. This acid rainwater then dissolves more limestone and the cycle continues. As chemical weathering of limestone accounts for more of the total chemical erosion on land than any other rock, this is of global importance. Weathering of limestone in mountains, such as the Himalayas, is thought to have a significant impact on the carbon dioxide content of the atmosphere, with consequences for global climate (see p. 101).

RIGHT Rock fragments produced by frost shattering form talus slopes on the sides of Redcloud Peak, Colorado, USA.

BELOW Exfoliated boulder with typical 'onion skin' pattern of weathering.

PHYSICAL WEATHERING

Chemical weathering predominates where temperatures and humidity are sufficient to encourage the chemical reactions that rely on water. But in freezing mountain and polar regions, or in the dry deserts of the world, or where there are few living things, physical disintegration is a much more important factor in determining the weathering of surface rocks.

One of the most efficient physical weathering processes is caused by the rapid freezing and thawing of ice. Water invades the joints or even tiny cracks in a rock, then freezes and thaws, freezes and thaws. The expansion of the water on freezing causes the ice to act like a wedge being driven into the rock, prizing it apart. In high mountain regions and in cold regions the debris of this physical attack lies on slopes as scree or talus. The growth of salt crystals in arid areas is thought to have a similar effect, expanding as they crystallise to crack rock. This can lead to a form of weathering (in basalts and granites) called exfoliation, in which the rock layers spall off from the surface like layers of an onion and may create 'onion-skin boulders'. Because these boulders are typically found in hot, dry lands, the phenomenon was previously thought to be due to the effect of repeated cycles of intense heat and cold, stressing and eventually fracturing the rock. (We do know that violent heat can fracture rock; Hannibal used fire to break boulders on his way across the Alps and, in ancient Zimbabwe, quarrymen used fire and cold water shock to break rocks.)

Releasing the pressure on rocks can also result in exfoliation. As rocks which formed at great depths become exposed through the erosion of other rocks above, they crack open, rather as though their 'belts' were being loosened and they were allowed to breathe out. Such pressure release causes joints in granite masses, and results in the spalling of thin sheets of rock from boulders and cliffs, which can occur on a very large scale.

LEFT Plant power – growing tree roots can exert enough force to crack a rock.

The more a rock's surface is physically cracked and weakened, the greater is the surface area open to chemical attack. And the more intense the chemical attack, the weaker and more vulnerable the rock is to physical break-up. Even in the driest desert, some moisture is present and will eventually do its work.

The speed and type of weathering is determined by the parent material, climate, topography, organic activity, and time. Weathering rates vary enormously across climatic zones, and operate on different scales. For example, in temperate regions granites are generally seen as stable, as the climate is relatively cold and the chemical changes proceed slowly. But in the tropics it is a different matter. Here, humidity and high temperatures speed up chemical reactions dramatically, and granite boulders in the ground are so rapidly weathered that they can often easily be kicked to dust.

Weathering is best seen as a process of transformation. When granite weathers to sand, clay and dissolved material, the process of change destroys one form of rock but in doing so creates the materials to make others. Vast quantities of clay and silt are produced to feed mud to the sedimentary system, as is quartz to form sands and sandstones.

ABOVE Soil's raw materials – living matter, gradually decayed, and the products of rock weathering.

SOIL

Perhaps the most important product of weathering for us is soil, which sustains almost all life on land, either directly, in supporting plant life, or indirectly, through the organisms that feed on it. Soil is composed of rock fragments – the products of weathering – and of decaying organic matter – the remains of plants and animals, broken down by insects and hosts of unseen soil-living micro-organisms. Soil is both the product of weathering and the site of further, intense weathering. When soils form, a positive feedback loop exists - the build-up of soil promotes the ideal conditions for further chemical weathering. This is because soil holds and concentrates the water and chemicals that continue the process of decay. Soils may be ten times as acidic as rainwater, because they include solutions of carbon dioxide from respiring bacteria and plant roots, and of other organic acids secreted by bacteria. Therefore a covering of soil over rock means that the rock is effectively bathed in acid. So, once soil starts to form, rock weathers more rapidly, and more soil forms.

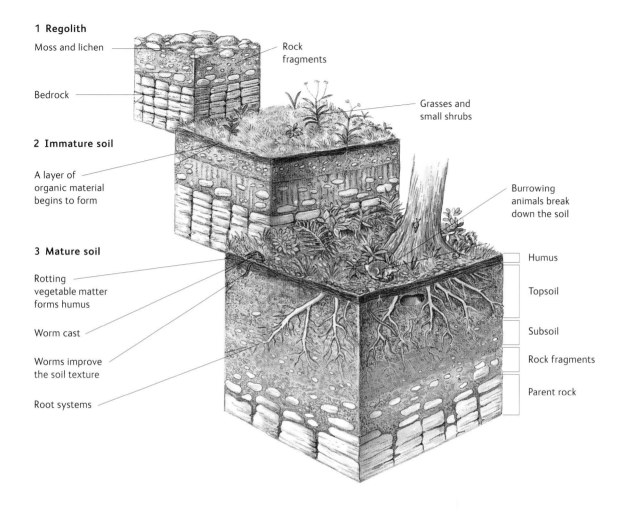

1 Regolith

Moss and lichen

Rock fragments

Bedrock

Grasses and small shrubs

2 Immature soil

A layer of organic material begins to form

Burrowing animals break down the soil

3 Mature soil

Rotting vegetable matter forms humus

Humus

Topsoil

Worm cast

Subsoil

Worms improve the soil texture

Rock fragments

Root systems

Parent rock

A single kilogram of soil may be home to an unimaginable number of living organisms. The vast majority are bacteria, fungi and millions of other assorted micro-organisms, and there are thousands of mites, insects, worms, and other creatures. Their activity is vital to the health of the soil. Charles Darwin first highlighted the role of earthworms in turning and aerating the soil, allowing more efficient breakdown and cycling of materials. (He calculated that the earthworms in a typical English garden could mix 6½ tonnes of soil per hectare per year.)

The activity of soil micro-organisms releases elements for use as nutrients by other living things. And chemicals released by weathering may be taken up or transported in solution to be deposited elsewhere, perhaps as cement to bind other rock fragments together, or as crystalline deposits in arid conditions. The decay process that happens in soil is thus a vital part of the cycle of chemical exchange between living and non-living domains, between water, air, rock and life. Because of this, we should see soil as part of the surface system, its own complex system being part of the Earth's larger interacting systems of exchange.

ABOVE Soil profile – soils are composed not just of the fertile top zone in which plants grow, but have a complex layered structure. Cutting a pit 1–3 m (3–10 ft) down into the soil reveals a soil profile, with layers called soil horizons. These have developed over time, and are the product of plant decay by animals and microbes, mixing by animals and the action of water carrying solutes percolating down to lower levels.

The way a soil develops depends on many things, but primarily on climate: temperature and rainfall are key factors. Higher rainfall and high temperatures activate more intense chemical weathering and biological activity. In arid climates, with little water, poor, thin soils develop. In warm, humid climates, deep soils may form, but these may not be as fertile as those of cooler climates. The tropical soils that support the lushest vegetation on Earth are surprisingly unproductive for crop plants. In the heat and humidity, nutrients released from decaying vegetation are quickly recycled in new growth. Also, the heavy rainfall washes rapidly through the soil and carries dissolved material down with it in a process called leaching. The few nutrients which remain are quickly exhausted by crops, so that in a few years the soil is of no value. Without any input from decaying vegetation it will not be renewed. And without vegetation, it is easily eroded.

VEGETATION ANCHORS THE LAND

Growing plants anchor soil and help to build it up. Coastal sand dunes, for example, build up around nuclei of specially adapted grasses such as marram grass which can survive the hostile conditions. And similarly hardy plants survive and grow in tiny pockets of little more than city grime, gradually accumulating sufficient soil to sustain themselves.

Vegetation is also an important protection against erosion. It reduces wind speeds near the ground and, together with moisture, it tends to bind surface particles together. In arid regions, and over poorly farmed fields stripped of vegetation, erosion may outstrip the formation of new soil. The infamous 'Dustbowl' of 1930s America was the result of intensive farming during the early 1900s in some southwestern states, which damaged and exposed the soil. Intense winds stripped the topsoil, blowing dust more than 1000 km (621 miles) eastwards and ruining over 35,000,000,000 m² (3½ million hectares) of farming land. Today, millions of tonnes of topsoil are still lost each year in the USA, at twice the rate of soil formation.

Erosion is also a worrying development of increasing tourism in fragile landscapes. Coastal paths, hillsides and even solid rocks are vulnerable. Modern footwear only adds to the problem as increasingly efficient grip disrupts soil cover more effectively. A single walker does little damage, but millions of footsteps put considerable pressure on the system.

OPPOSITE The mangrove swamps along the Atlantic coast, east of the Amazon estuary, Brazil. The roots of the mangroves anchor in the sediment, and allow sediment to build up around them.

ABOVE A severe case of erosion on a well-trodden path in Derbyshire, UK.

EROSION

A S WEATHERING SLOWLY WEAKENS and breaks down rocks at the Earth's surface, the more active processes of erosion and transportation take place. Weathering and erosion operate together to lower the surface of the land, and their combined effect is known as denudation. This process sculpts the landscape and releases vast quantities of material into the global systems for recycling. Today, we are only beginning to quantify the extent of this process on a global scale.

OPPOSITE The extraordinary beauty of Antelope Canyon, Arizona, USA owes its form to erosive action of both wind and water.

AGENTS OF EROSION

Gravity is a major source of power that results in erosion on land. Any material on a slope – rock, soil, water or ice – has gravitational potential energy. As it begins to move downwards, this potential energy is transferred into movement – kinetic energy. In 'mass movements' such as rock falls and landslides it is the debris produced by weathering itself which moves under the influence of gravity. With rivers and glaciers, gravity provides the energy for water and ice to flow and this flow in turn provides the energy to erode and transport material from the land's surface. When ice and water flow they seek the steepest and shortest route downslope, ultimately to the sea. This provides the most rapid conversion of gravitational potential energy to kinetic energy.

Water is probably the single most important agent in shaping the form of the land surface. Flowing water has sufficient power to pluck and sweep away soil, grains and pebbles, and even when in flood, to lift and hurl great boulders. Over time, it cuts deep valleys down into the land. And, all the time, it abrades and changes the shape of the surface over which it moves. Waves too can exert enormous power – a wave 10 m (33 ft) high can strike with four times the energy of the three main orbiter engines of the space shuttle.

Ice can also be an important erosional agent. If snow becomes permanent, it becomes compacted and recrystallises to form glacial ice – a crystalline 'rock' made of the mineral ice. As this ice accumulates, it begins to flow under the influence of gravity like a very viscous liquid. The individual molecules of ice slide across one another like cards in a pack, and the whole mass creeps slowly downslope, bulldozing everything in its path. The huge mass of ice has power to crush and gouge rocks and to carry with it boulders the size of houses. Rock fragments freeze into the base of the glacier and are plucked up and transported. Finer rock debris acts like sandpaper, scratching and scraping the underlying rock.

Wind erosion is a more selective process. It acts effectively when there is no moisture or vegetation to bind loose sediment, and is therefore only a major erosive agent in dry regions of the world. Because air is much less dense than water, wind can usually move only small rock particles such as sand grains. Larger particles tend to be left behind in a selective process called winnowing. Gradually, these fragments form a gravelly 'desert pavement' protecting the layers of finer material underneath.

The world's hot deserts are a major source of atmospheric dust. An estimated 130–800 million tonnes is blown from the surface of continents annually, most removed in great dust storms. The Sahara has been estimated to lose between 60 million and 200 million tonnes per year. Fine particles can be lifted and carried in suspension in the atmosphere for thousands of miles and for many years. (Studies of drill cores from the oceans, where most dust ultimately settles, have helped to establish the sheer volume and effect of windblown erosion on a global scale and the significant contribution of windblown deposits to ocean sediments.)

The other main energy source responsible for erosion and redistribution of material around the globe comes ultimately from the sun, in driving currents of wind and water, and from the gravitational attraction of both sun and moon which together drive the tides (see p.16). The stronger the wind and the greater the expanse of water over which it blows, the larger the waves that buffet the coastlines they encounter. And, as tides rise and fall, water moves in and out from the shore as a broad sheet to erode and redeposit beaches of sand and shingle.

RIGHT Waterfall and river in the Grand Canyon, Yellowstone, USA.

ABRASION

The erosive power of ice, water and wind is greatly increased by the material they carry suspended with them. These particles behave like tools to wear away solid rock.

Slow moving ice drags sharp rock fragments with sufficient power to gouge grooves in boulders and hard bedrock. Finer fragments of crushed rock moving with the ice polish surfaces to a high sheen.

Armed with suspended sand and dust, wind will sandblast rock surfaces, sometimes forming pedestal rocks. These may form because a harder, resistant cap of rock lies above softer underlying rock, which has eroded away. Or the structure may mark the upper limit of the sand's abrasive action.

Ventifacts are wind-abraded pebbles created by sandblasting on the side facing into the wind. Occasionally, a sudden gust or storm will turn the pebbles, exposing another face to wear. Dreikanters are three-sided ventifacts ('Dreikanter' is German for three corners'). Their polished surfaces have been abraded by fine dust.

Loose rocks in a river rub against one another, smoothing and polishing. Sometimes, pebbles caught and swirled round in a depression in the streambed wear the rock to form a pothole.

BELOW Switzerland's Matterhorn – a classic frost-shattered peak eroded by glacial action.

EROSION AND LANDSCAPE

Erosive forces sculpt the land. The landscapes we now see around us were produced by past erosion, and their forms can indicate the nature of long-past environments. Our interpretation of these forms relies on our knowledge of processes active today – understanding the present is the key to understanding the past.

Glacial erosion leaves very characteristic marks on the landscape. Ice forming at the head of a mountain valley gradually erodes a hollow where the snow accumulates (called a cirque). If two glaciers form on either side of a rock divide, they gradually erode that divide away to form a sharp narrow ridge (called an arête). And if three or more are involved, a pinnacle or 'horn' may result, such as the dramatic Matterhorn in Switzerland. Moving glaciers gouge out great U-shaped valleys that may deepen enough to cut across valleys of tributary glaciers to leave these as hanging valleys.

In contrast to glaciers, rivers cut deep, V-shaped valleys through the rock, their natural cutting action being straight downwards, always seeking the fastest, least resistant route downhill. They can form spectacular erosional landscapes such as the eroded valley of the Yellowstone River which cuts through the national park in Wyoming, USA.

Over time, land may be uplifted, or relative sea levels may fall for other reasons, altering a river's gradient. This re-activates the river's erosive action and causes it to cut even further down to the new base level. The Nile, for example, once cut a canyon the size of the Grand Canyon, when the Mediterranean had dried up (see p.72) and the river had to cut much deeper before reaching the sea. The Rhône flowing into the Mediterranean at its western end also cut its own, even bigger canyon. Both of these canyons are now beneath the sea and have filled with sediment, so their presence can be detected only by seismic surveying beneath the sea floor.

While wind erosion can lead to very distinctive forms, as seen on the previous pages, not all desert landforms can be attributed to wind action. Water can still be a powerful agent in arid areas; rare sudden floods in the wet season can rapidly scour out valleys (wadis) that soon become dry as the season passes. It is at the coast that some of the most dramatic effects of water erosion can be seen, in the destructive

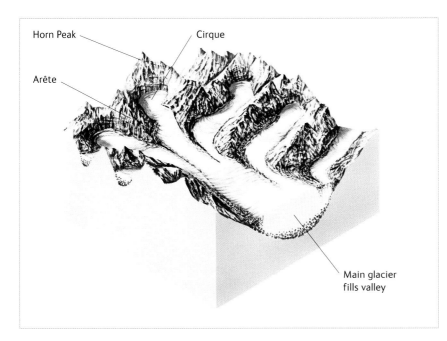

Horn Peak

Cirque

Arête

Main glacier
fills valley

Horn Peak

Cirque with lake

Arête

Hanging valley

U-shaped
valley

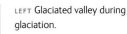

power of the waves. Waves crashing into a cliff-face gradually undercut its base, until eventually the overhanging rock collapses. Continued erosion causes the cliff-line to retreat leaving a wave-cut platform.

Any weakness in the rock is exploited. Blowholes can form where crashing waves force air under great pressure up through cracks in the rock, and on a larger scale whole coastlines of less resistant rocks are cut back to form bays and inlets. Caves form in more resistant headlands and are quarried through by the sea to form an arch; when an arch collapses it leaves isolated pillars called stacks.

RIGHT Stacks, formed by marine erosion. The Twelve Apostles along the Great Ocean Road, Australia, taken in December 2004 before the 'apostle' in the front had fallen.

UNDERWATER AVALANCHES

Erosion is not just a land event: avalanches and slides under the sea create deep submarine canyons much like their land counterparts. These occur along the edges of the continental shelves that border the oceans. Collapse of these margins can result in an underwater avalanche of sediment cascading chaotically downslope to the ocean plains. The dense mix of sediment and water set up by this is called a turbidity current. Eventually, the current wanes and the sediment it carries gradually settles as a turbidite layer with coarse grains at the base overlain by increasingly finer grains above.

Turbidity currents are usually triggered by a sudden erosive slump. One dramatic example of the force and speed of a turbidity current came in 1929, triggered by an earthquake off the Grand Banks of Newfoundland, USA. About an hour after the earthquake, a series of underwater telegraph cables snapped and could be tracked and timed. The current carried sediment for at least 700 km (435 miles) and reached speeds of 40–55 km/hour (25–35 miles/hour).

ABOVE Salt Point State Park, northern California. These are huge turbidite deposits, each layer representing an ancient underwater avalanche event.

SLOPES ON THE MOVE

The loose debris of weathering and erosion is far less stable than solid rock. Study any slope – whether high mountainside, sea cliff or gentle hill – and you will see the action of gravity: of fallen rocks, in fast mudflow and slow soil creep, and in the sudden slumping of whole tracts of land.

Mass wasting is the movement of material downslope under the influence of gravity – classically without the aid of water, ice or wind, although these may be involved in some way. In order for mass wasting to occur, a slope must become unstable. Like a car parked on a hill, when the hand-brake is released (friction is overcome), the potential energy of the car is converted into kinetic energy – movement. And the steeper the slope the greater the tendency to slide.

The tendency of material to move downhill is dependent on many factors other than the original steepness of the slope: the types, shapes and orientation of the rock fragments or layers, the nature of unconsolidated materials, the presence of water and/or vegetation and the action of earthquakes and volcanoes in triggering the mass-wasting event.

Water can have a dramatic effect on the stability of a slope. First, because water-saturated material is considerably heavier than when dry, it has greater potential energy under gravity. Water also lubricates, reducing friction. Finally, as water freezes and thaws, it alternately pushes apart and re-arranges the packing of particles, weakening their cohesion. Often, the action of water goes unnoticed, until the final collapse. The examples overleaf illustrate the consequences of such landslides, creating dams and lakes, causing overspills from dams and burying villages.

The impact of raindrops alone can erode an exposed sandy slope. Heavy rainfall can cut deep gullies into unstable slopes, or have more catastrophic effects. The water may form a slurry of mud and rocks (mudflow) that can sometimes travel in excess of 120 km (75 miles)/hour.

When a whole hillside becomes unstable and moves, a mass of land or rock may slump or fall in one sudden movement. Such slides are usually small-scale, in geological terms. However, there are spectacular exceptions. One of the largest landslides on Earth, in Saidmarreh, southwest Iran, happened 10,000 years ago. A mass of limestone 15 km (9 miles) long, 5 km (3 miles) wide and 300 m (984 ft) thick travelled 18 km (11 miles) downslope. The slide was caused by destabilisation of underlying sediments.

Loose material can move downhill as a viscous fluid mass, like road tar on a hot day. Such flows can be extremely rapid, but are usually very gradual. 'Creep' is the slow downhill movement of rock or soil, moving so slowly that it is hard to imagine it as fluid. Its action can be detected in ledges on hillsides. Because the surface material moves more rapidly than deeper levels, objects in the soil are tilted forwards. If you see a tree leaning on a steep slope or tilted gravestones in a hillside graveyard, this generally means that the soil is on the move.

Ridges across the slope, and trees and posts at angles indicate soil creep

Large 'steps' in the hillside are evidence of slump

Mudflows occur when water lubricates a mass of loose particles, making them flow as one mass

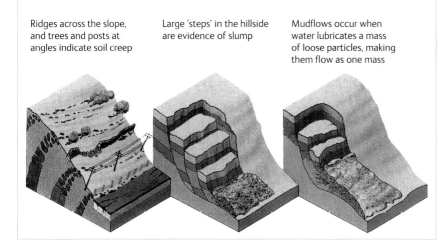

LEFT Basic patterns of movement downslope: soil creep, slump and mudflow.

LANDSLIDES

Eroded material may sometimes flow, slide or fall with astonishing speed, and with terrifying impact on human life. We can't predict the unpredictable. But if we fail to take proper account of how material behaves on a slope, we may court disaster.

Aerial view of the devastation caused by the volcanic eruption in Nevada del Ruiz, Armero near Bogota, Colombia in 1985. A vast mudflow of water, ash and debris raced down gullies and stream valleys to the town of Armero 48 km away from the volcano.

The landslide, causing a veranda of a house to hang over the cliff edge, after the El Niño storm in California, USA, 1998.

A landslide down a mountain in Guinsaugon, central Philippines after a night of heavy rain in February 2006 buried an entire village.

The 12 May 2008 Sichuan earthquake caused a landslide which created a dam resulting in the Tangjiashan lake. People were evacuated from the path of the potential floodwaters building up behind the dam, as a channel was carved out to try to drain away the waters.

The valley in the Italian Alps where 26,000 died in 1963 when 250 million cubic metres (8829 million cubic feet) of rock slid into a reservoir causing a large wave to spill out of the reservoir dam and drown villages downstream.

TRANSPORT

Across the world, vast quantities of matter are travelling, carried in the flow of ice, water and wind. More than 20 billion tonnes of solid sediment particles are transported to the oceans by rivers, annually. About the same mass is carried in solution. Further millions of tonnes of dust are carried in global wind currents, sometimes for thousands of kilometres. (Dust from central Asia has been found 11,000 km (6835 miles) away on the island of Hawaii.)

Each medium – ice, wind, water – has a different capacity to carry material, and this depends on many factors: the speed and steadiness of the flow, the viscosity of the medium, the relative density of medium and particle, and the sizes and shapes of the particles themselves.

BELOW A sudden flooding changes the landscape of Tsauchals River, Namibia.

Compared to feathers and fine dust, for example, air is a viscous fluid and so these things only sink slowly in air. Wind carries them easily, but even with the force of a tornado, it cannot carry a heavy mass. Storm floods and giant waves, by contrast, can sweep away structures as large as houses, and ice has the power to hold and carry great boulders for miles. (This is how erratics occur – see p.54)

The properties that most strongly influence the flow of these media are density, viscosity and velocity. The less dense a medium is in relation to a particle, the harder it is for it to hold the particle suspended against gravity. Viscosity determines the ability of the fluid to flow. It is the result of the interaction of molecules or particles within the fluid. Viscosity increases as the friction between molecules increases, preventing them from easily moving past each other. (The simplest way to understand the effect of this resistance to movement is to imagine the difference between running in open air, through water, and wading through mud.) As fluid velocity increases, there comes a point when sediment transport can occur, overcoming the opposing forces of gravity and cohesion.

Particles can move along with a fluid in three different ways: rolling, bouncing or by being swept up into the flow. For any given velocity, where the heavier particles are rolled along, lighter particles move down current in hops (a process called saltation) and the lightest particles are borne along in suspension by the current.

Wind usually lifts only the lightest grains and dust. Turbulent air currents over deserts may transport fine sand particles, while coarser sand particles are transported by bouncing, rolling and sliding along the surface. Heavier grains may saltate along the ground, quickly falling back down to earth under gravity, while pebbles and rocks are left behind.

Water plunges downhill in the youthful upper reaches of a river, abrading and transporting its load of pebbles. Moving more sedately on the gentler, lower slopes it meanders across open valleys carrying with it suspended silt and sand. The turbulence of water lifts and holds small grains in the flow, while larger grains and pebbles tend to be dragged along the bed, moving by small hops.

BELOW Particle movements in water.

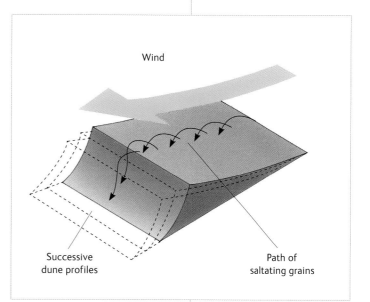

Wind

Successive dune profiles

Path of saltating grains

ABOVE On the surface of sand dunes, grains are usually on the move, causing the whole dune to migrate slowly forward. The wind drives grains of sand to the top of the dune, where they build up and then tumble over the top.

Rivers swollen in flood are commonly very cloudy with sediment, while slow-moving streams can be crystal clear, their sediment having all settled out.

Ice is so viscous in its flow that it can be considered a solid, but one that still has capacity to flow. Material incorporated into the base and body of the ice is given a 'free ride', held in suspension in the ice until melting releases it. Huge boulders can be carried as easily as fine grains. On melting, the water released from the glacial ice continues to transport some of the finer material, forming 'outwash' deposits.

CHANGED BY TRANSPORT

Once rock material is eroded, this is not the end of its transformation – whether it is cobbles, small pebbles or sandgrains buffeted in the wind. Environmental conditions and length of transport both affect continuing changes. Resistant minerals such as quartz are less likely to be destroyed by this abrasion, so they become predominant in the eroded load as it becomes more 'mature'.

Imagine sharp talus fragments fall onto a glacier. Frozen into the ice-mass, there is no abrasion between them, and little chemical weathering occurs at such low temperatures. The fragments remain poorly sorted, angular and immature. They will reach the foot of the glacier relatively unchanged.

Imagine next that the talus is swept further downslope by meltwater to join a river. The fragments are rolled and bounced along to round and smooth their jagged edges. Meanwhile, the chemical action of water gradually transforms them, breaking down and dissolving some constituents until only the most insoluble material is left.

ABOVE As particles are transported they tend to be rounded by abrasion and separated by size to give a well-sorted sediment like this.

LEFT A dust storm in Iraq lifts and carries millions of tonnes of sand and finer particles.

DEPOSITION

W ORN AND TRANSPORTED BY WIND, ice and water, rock and mineral fragments eventually settle. Deposition occurs because the transporting medium no longer has the energy to carry the material. Ice melts, winds die down, water flow slows, and their burdens of sediment begin to fall under gravity.

Along its course, a river acts like a long-distance sieve, sorting the material it carries as it goes. Wherever it slows, heavier rocks and pebbles are left behind, while finer particles may travel onwards in the flow. Where the river flows into the sea, the far larger volume of seawater dissipates the river's energy and the river sheds its load,

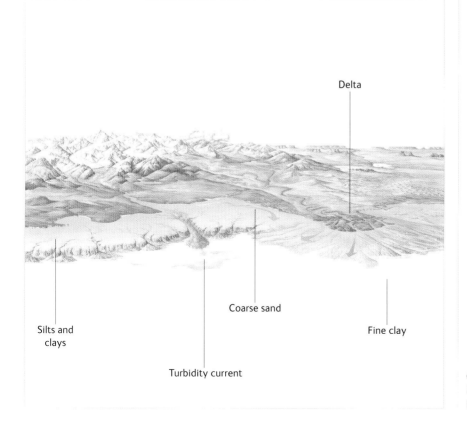

Delta

Silts and clays

Turbidity current

Coarse sand

Fine clay

LEFT A delta builds up and out where a river enters the sea. Further offshore a turbidity current carries sediment downslope along the seafloor.

OPPOSITE A satellite image of the Mississippi delta, USA, clearly shows its 'bird's foot' form.

again in sequence. Coarser grains drop close to the river mouth, while silts and clays are carried on with the weakening current to settle further out to sea. Most coarse sands will be deposited within tens of metres of the coastline but may then be distributed much further by seabed currents. Fine-grain sediments will be deposited to form muds further offshore with the finest components being carried to the deep ocean.

At the margins of the world's great landmasses lie billions of tonnes of such land-derived sediments. The great majority of these have been carried by rivers to the sea. Some are carried beyond the edge of the continental shelf by slides, slumps and turbidity currents (see p.42). Submarine exploration and the study of deep-sea drill-core samples has made it possible to estimate the vast scale of this transfer of material from land to sea. Submarine fans of turbidite deposits larger than the size of Britain exist off the Mississippi and St. Lawrence Rivers. In the Indian Ocean, there is a massive shift of material from the Himalayas in particular to the seafloor.

Where a river runs into the sea or lake it may deposit sediment in the form of a delta. Deltas develop different shapes depending on the balance between deposition and erosion. As fans of sediment build out and up, the river may break through its banks and form new routes to the sea. The Mississippi River has changed its course seven times over the last 6,000 years. Its delta, above and below sea level, now extends about 1,600 km (994 miles) beyond the main coastline, forming a 'bird's foot' delta and depositing an estimated two million tonnes of sand and silt a day.

Throughout history, societies have relied on the fertile soils of floodplains and deltas to nourish their crops. The people of Bangladesh, for example, live in precarious balance with the rivers that both supply and destroy the land on which they depend. The floods often cause untold damage and misery, but without them Bangladesh would not exist. Modern developments such as the building of large dams can upset this ancient relationship between people and nature. The Aswan dam of the Nile, for example,

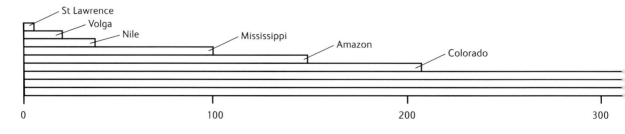

St Lawrence
Volga
Nile
Mississippi
Amazon
Colorado

0   100   200   300

traps vast amounts of sediment. This starves the river of sediment downstream, and is leading to the erosion of the existing fertile delta. Heavy sedimentation also reduces the storage capacity of the reservoir and damages the dam's machinery. Many of the world's large dams are suffering similar problems, caused by human failure to take proper account of the natural processes of transport and deposition.

ABOVE Housing destroyed by flooding caused by heavy rain.

PLACES TO SETTLE

Sediment settles out when energy levels fall in the transporting medium's flow. This may not always be at the coast; other major sediment traps exist on land – and far out in the deep sea. It may take hundreds or even thousands of years to move this stored sediment elsewhere.

BELOW Three great rivers flow off the Himalayas and Tibetan Plateau (Ganges, Huang He and Brahmaputra) and carry between them 20% of the world's land-derived sediment. (Other rivers shown for comparison.)

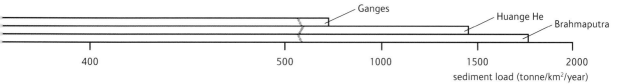

Ganges

Huange He

Brahmaputra

| 400 | 500 | 1000 | 1500 | 2000 |

sediment load (tonne/km²/year)

LEFT BY GLACIERS

Besides picking up material by erosion, glaciers also carry fallen rock debris from higher mountains and valley sides. This material is known as moraine, and it can be seen as great streaks darkening the glacial ice. When the supporting ice melts, it will be dumped to form a mound or ridge. The deposit itself, called till, is an unsorted mix of fragments, from boulders to pebbles and sand. (It was his observations, on the close similarities between moraines of modern glacial environments with older deposits and rocks over northern Europe, that led geologist Louis Agassiz to propose in 1840 that the northern hemisphere had once been covered by giant ice sheets.)

In areas that have been glaciated, large boulders are occasionally found lying free on rock surfaces from which they were not derived. Boulders of this type are called erratics and may have been transported thousands

ABOVE An erratic.

of kilometres by the moving ice. As a glacier melts and retreats, streams flow out from within and beneath the ice carrying the finer-grade material with them. This material will gradually be sorted in the water's flow and deposited elsewhere (see, for example, 'Loess' on p. 57).

A lake delta formed at the mouth of the River Maggia, Switzerland.

Alluvial fan marking the break of a slope.

LAKES

When a river flows into a lake (or artificial reservoir) it will dissipate its energy on entering this large, still volume of water, and sediment settles out. In sufficiently large lakes, deltas may form.

ALLUVIAL FANS

When a stream laden with sediment emerges from a steep mountain gorge down to a wide-open plain, it drops its sediment as the velocity and turbulence of the flow decreases rapidly. Gradually a cone of sediment called an alluvial fan builds up at the foot of the mountain.

FLOODPLAINS

Floodplain deposits arise when a river in a flat-bottomed valley bursts its banks and flows over the adjacent plain, depositing fine sediment over the flat land. These fine silts and clay deposits help produce good, fertile land for agriculture.

DESERTS AND DUNES

In extremely dry areas where there are no rivers to remove sediment and transport it to the sea, sediment accumulates as vast seas of sand known as ergs, the classic deserts most people are familiar with. Ergs may cover as much as 500,000 km^2

BELOW A river meanders through its floodplain.

(310,686 miles^2), as in Saudi Arabia. Sand dunes may be only temporary sediment stores, as the wind continues to pick up and redeposit the sand, and sometimes carry it far away from the desert region. They tend to migrate downwind because sand is constantly moving up the shallow upwind side and being tipped over the steep downwind slope (see bottom images p.47).

LOESS

Vast expanses of accumulated, wind-blown dust and silt called loess can be found across the northern hemisphere. This sediment probably originated at the margins

BELOW The sand dunes of Erg Chebbi in the Sahara Desert near Merzouga, Morocco.

RIGHT Loess deposits in Gansu province, in the upper basin of the Yellow River (Huang He). The upper slopes are almost bare soil dissected by stream gullies in what is the natural landscape of loess hills. The lower slopes have been terraced to extend the cultivable area of the valley floor.

OPPOSITE This eroded cliff-face in the Colorado National Monument, USA shows hard and soft layers of sedimentary rocks.

of retreating ice sheets. High winds blowing off the ice sheet picked up the fine outwash sediment and redeposited it further away as glacial loess. Loess can be carried large distances in suspension. It is estimated that 10% of the world's land surface is covered by loess and loess-like deposits which form extremely fertile soils, for example in China and Eastern Europe.

DEEP SEA

Fine-grained red clays are found in the deep parts of the seas. They accumulate very slowly at less than 1 mm (less than 1/100 inch) every thousand years. These clays contain very fine particles that have been carried in suspension far from the land or blown out to sea.

BELOW Life on the abyssal sea floor, around a depth of 4000–6000 m (13,123–19,685 ft), near the Hudson Canyon off the coast of New Jersey, USA.

LAYERS, BEDS AND OTHER PATTERNS

Beds (layers of sediment) are perhaps the most obvious feature of sedimentary rocks, especially when seen exposed in cross-section at a cliff-face. As deposits of soft sediment build up, they settle in a series of layers. Each is called a bed, and the junctions between them are called bedding planes. Series of thin layers, only a few millimetres thick, are called laminae, but when the layer is many metres thick, the bed is said to be massive. Slight variations in the composition and structure of the layer means that they may be differentially eroded, to produce a cliff-face that looks like a giant stack of boards.

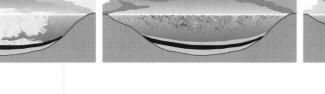

ABOVE Varve formation: a spring flush of sediment carried by glacial meltwaters in the spring is followed by a summer bloom of phytoplankton. Dying plankton darken the sediment.

Sedimentary layers are geological records of the passing of time, and of environmental change. Most of us are familiar with the rings of a deciduous tree trunk, the concentric layers which mark the annual rhythm of summer and winter growth, and there are sediments which also record time in their layers. An example is the varve layering of sediments, typical of lakes fed by glacial meltwaters. In spring, the meltwater flushes a coarse-grained, light grey clay sediment into the lake, where it settles. Nutrients brought into the lake, and the increasing warmth, promote the growth of microscopic plants called phytoplankton. As the season progresses, these create a 'bloom', then die and their microscopic bodies drift to the bottom of the lake to darken the sediment. Year after year, the process is repeated, to create the layers of the varve deposit.

If we make the basic 'uniformitarian' assumption that layers are laid down horizontally and in order, so that the lowest layers are the oldest, then layers can tell a sequential story of change over time. The sequence may be simple, as in the varve, or it may be complex, telling the story of major environmental or climatic changes. The sedimentary layers of the Grand Canyon, for example, record 500 million years of the evolution of life in the fossils preserved in its sequence of layers.

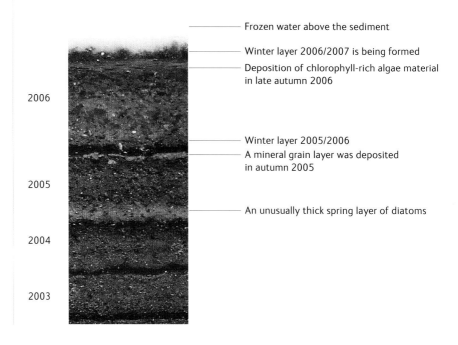

Frozen water above the sediment

Winter layer 2006/2007 is being formed

Deposition of chlorophyll-rich algae material in late autumn 2006

2006

Winter layer 2005/2006
A mineral grain layer was deposited in autumn 2005

2005

An unusually thick spring layer of diatoms

2004

2003

RIGHT Varves deposited in a lake in northern Sweden between 2003 and 2006.

PATTERNS IN THE SEDIMENT

Sedimentary layers often show features that can help in identifying how the sediment was transported and deposited. The most obvious is a sequence of horizontal layers formed by episodes of sediment settling.

Flowing water or wind can sometimes mould the loose sediment into ripples. A current flowing in one direction creates *asymmetrical* ripples, while a tidal current (oscillating backward and forward) forms *symmetrical* ripples (BELOW).

Other, usually transient structures can also be found – marks as delicate as rain spots, the tracks of a skipping stone or the swishing of a seaweed frond across soft mud. Animals that live on and in the sediment may leave their tracks and burrows. Normally, these would disappear as swiftly as a footprint in the sand when the tide comes in. But, very

occasionally, new sediment fills the dips and hollows they have made and they are preserved as 'casts' in the hardened rock.

In this specimen (ABOVE), a layer of dried-out mud was overlain with darker sediment, which filled in between the cracks. The whole deposit was then compacted and changed to rock, 'fossilising' the pattern.

A special pattern called graded bedding is the result of a sediment of mixed grain sizes settling through water. The smaller, lighter and flaky particles encounter more resistance to their fall through water. The heavier particles reach the bottom first, therefore, and the finer grains settle on top of them to form a sequence that becomes finer upwards – a graded bed (BELOW).

READING THE CHANGING PATTERN

Sedimentary rocks, being the products of surface processes, are a powerful means of reconstructing geological history, and also of understanding the processes and patterns of change. The previous page gave examples of how individual layers and beds can be read to discover how the rocks came to be formed. Series of layers can also record a gradual change over longer periods of time, for example, in the silting up of an estuary to form land.

Sometimes, however, rock sequences preserve evidence of more dramatic changes that took place in the intervening time between the deposition of different beds of sediment. These are called unconformities. An unconformity represents a time-gap in the geological record. It marks a distinct 'break' in a sedimentary sequence where deposits have been removed by erosion prior to new sediments being deposited. The 'missing' rocks represented by the unconformity have been

RIGHT An unconformity is a break between two rock types laid down at different times. The image here from Kilkenny Bay, north Somerset, UK, shows an angular unconformity (or nonconformity), in which the rock layers are at different angles. Here sedimentary rocks (upper half) were formed by the deposition of successive layers of particles which became compressed. This occurred after the underlying igneous rock was folded.

eroded away over thousands or even millions of years before the next layer of sediment was laid down.

It was an unconformity that influenced James Hutton to argue for events of long-time change on Earth. One unconformity he studied consists of a series of parallel sedimentary beds lying at an angle beneath a further series of horizontal layers. Hutton was one of the first geologists to realise the implications presented by such field evidence. He proposed that the first layers were sediments laid down in a shallow sea. After being buried deep and hardened to rock, these formations were then tilted, folded and uplifted to the surface, where they were eroded. Much material was removed,

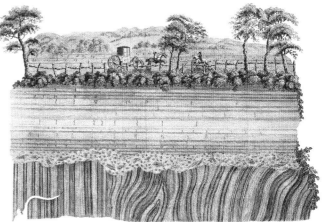

ABOVE Engraving of the unconformity used to illustrate Hutton's original study.

levelling these rocks until the sea covered them once more. New sedimentary layers were deposited unconformably on the much older rocks. Then, the layers were uplifted – to be eroded again. And so, for Hutton, the cycle would continue over immense periods of geological time.

Today, our knowledge of what causes such changes from sedimentary to erosional environments is greatly helped by our knowledge of climate change, and of plate tectonics. Tectonic movements can thrust rocks upwards from deep levels in the Earth's crust and pile them up into mountains – most spectacularly, today, in the peaks of the Himalayas, Rockies, Andes and Alps. Moving plates also change the pattern and volume of the oceans, and so affect sea levels over geological time. In coastal regions, sedimentary layers may emerge above sea level and become eroded, or submerged below the waves to be overlain with new sediment.

The relative level of land to sea can change over time for a number of reasons. Variations in global climate affect both the thermal expansion of water and the amount of water locked up in ice. (Global change in sea level is called eustasy.) In the last Ice Ages, so much water was locked up in the ice sheets that the global sea levels fell more than 100 m (328 ft). Huge deposits of sediment extending from the shoreline on shallow continental shelves were affected most severely by the large-scale changes in sea level associated with this period of glaciation. They were exposed and eroded, as rivers cut down deep valleys to reach the new, lower sea level. As the seas returned, these valleys were flooded again.

RIGHT Raised beaches can be caused by isostatic readjustment. When ice weighs down the land, a new beach is cut at this level. When the ice melts, the land rebounds and this beach is exposed above the present sea level.

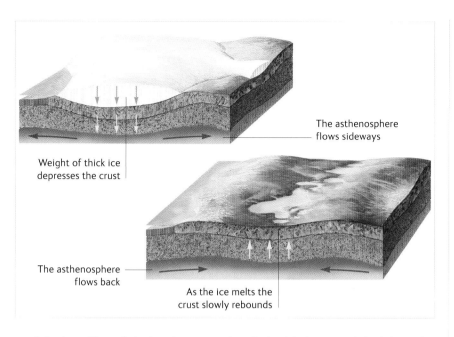

The asthenosphere flows sideways

Weight of thick ice depresses the crust

The asthenosphere flows back

As the ice melts the crust slowly rebounds

LEFT During the last Ice Age, vast ice sheets weighed down parts of the Earth's crust. Today, the land is still rebounding. This process of adjustment is called isostasy.

A further effect of the ice sheets was that their added mass weighed down the continental lithosphere, again altering the sea level. The lithosphere sits on a layer of the Earth's mantle called the asthenosphere which is hot and has a plastic, flow-like behaviour. The heavier the continent, the more it sinks into the mantle. Remove the weight and the continent rebounds. (Scandinavia is still rebounding by as much as 2 cm a year from being pressed down by the ice.) This property of readjustment is called isostasy.

FROM SEDIMENT TO ROCK

S EDIMENTS ARE CONSTANTLY SHIFTING, unstable accumulations. But if they are buried, they may become solid rock through a series of physical and chemical changes. Mud, made up of silt and clay, will turn to mudstone or shale, sand to sandstone and gravel to conglomerate. The time taken for this to happen varies widely – some sediments remain loose for tens of millions of years.

To turn sediment to rock, two main processes must occur: compaction and cementation. The major physical change is compaction, when the mineral grains of a sediment are squeezed together by the weight of further overlying sediment. Different sediments compact by different amounts. If you look at a sandstone, its structure seems very little different from loose sand, because sand deposits are already fairly well packed. Newly deposited silt and clay deposits, however, may contain as much as 90% water and their volume reduces drastically as this water is squeezed out. The other major processes involved in turning sediment to rock are usually chemical. Water seeping through sediment carries with it material dissolved from the surface or from the minerals it flows over. As conditions such as temperature and pressure change, this dissolved material can be deposited as cement to bind the particles of

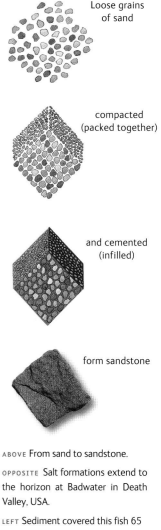

Loose grains of sand

compacted (packed together)

and cemented (infilled)

form sandstone

ABOVE From sand to sandstone.

OPPOSITE Salt formations extend to the horizon at Badwater in Death Valley, USA.

LEFT Sediment covered this fish 65 million years ago. The sediment then gradually turned to rock, preserving the fish.

ABOVE **Close up of a shale cliff.**

sediment together. The commonest cements are calcite, quartz and iron oxides (the red cement that colours red sandstone). Cementation can also occur when *physical* pressure causes the particles of sediment to dissolve at points of contact. When the dissolved material recrystallises, it will help cement the grains together.

On cementation, further chemical changes may also occur to alter the mineral composition of the rock. In most sediments, the processes of weathering, transport and settling have brought together minerals from different sources. Now, buried and pressed together in a new environment, the mix of minerals gradually readjusts. Some of the original minerals may dissolve and new ones may be precipitated. Occasionally, there will be localised changes. Minerals may crystallise around a 'seed' such as the remains of an animal or plant, resulting in distinctive concretions that can be easily separated from the surrounding rock (for example chert nodules in limestone).

So far, this book has largely been concerned with the erosion, transport and settling of rock fragments to form sedimentary rocks. But there are other sources of material, and other types of sedimentary rocks.

LEFT A millstone grit outcrop at Stanage Edge, Derbyshire, UK.

LEFT Large ooliths bound together by a cement of calcite.

THE ROCK CYCLE

A sedimentary rock is just one stage in a continuing process of change and exchange at the surface of the Earth, and between surface and interior. Hutton proposed, over 200 years ago, that the surface underwent an unending succession of uplift, erosion, sedimentation, burial and melting.

This 'rock cycle' is essentially a grand geochemical cycle, as minerals are transformed and the chemical constituents of some are redistributed to form new mineral suites in different varieties of rocks. Material is also exchanged between the solid crust and other surface systems – the atmosphere, hydrosphere and biosphere.

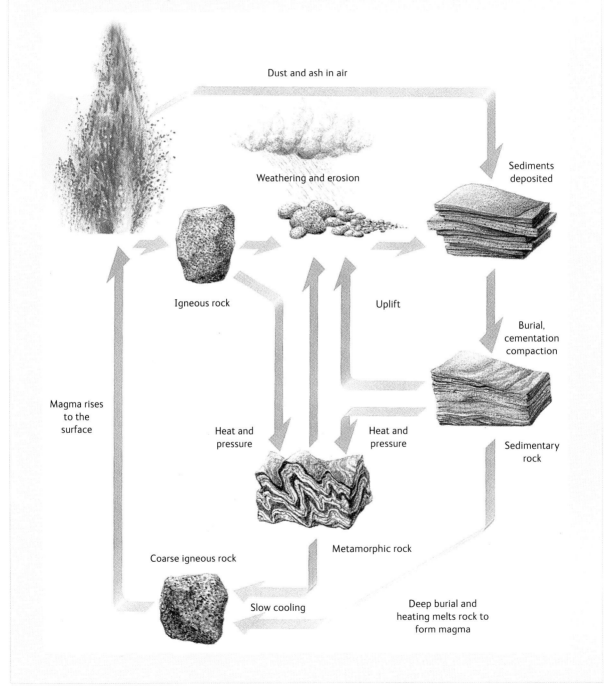

Dust and ash in air

Weathering and erosion

Sediments deposited

Igneous rock

Uplift

Burial, cementation compaction

Magma rises to the surface

Heat and pressure

Heat and pressure

Sedimentary rock

Coarse igneous rock

Metamorphic rock

Slow cooling

Deep burial and heating melts rock to form magma

IN AND OUT OF SOLUTION

Natural water is never 'pure', it contains dissolved chemicals that were once part of a rock. Carbon dioxide dissolves in water to make it slightly acidic (described on p.25) and this increases the water's ability to react with rocks and soils, dissolving the minerals they contain, releasing their constituent atoms into solution as ions.

Water flowing over and through the surface of the land carries away these ions – this is the river's dissolved or solute load. The world's rivers carry about 20 billion tonnes of solute a year. The dissolved material carried to the sea by the rivers are often characteristic enough for us to detect their 'signature' many kilometres from land. The offshore waters of the Indian Ocean, fed by the rivers running off the Himalayas, have a different chemical composition from those of the Atlantic where the Amazon disgorges. Each river has picked up its own special blend of dissolved ions on its journey through different geological terrains.

The sea is saltier than the rivers that feed it, but not as salty as one might expect if this influx of dissolved material ended with its entry into the oceans. Records from ancient deep-sea deposits suggest that the sea's salinity has been relatively stable for much of the Earth's history. Therefore, billions of tonnes of dissolved ions must somehow be removed to maintain this balance. Material is removed by precipitation of crystals out of solution and by incorporation into the bodies of

BELOW Water cycle – water is constantly recycled between sea, atmosphere and land.

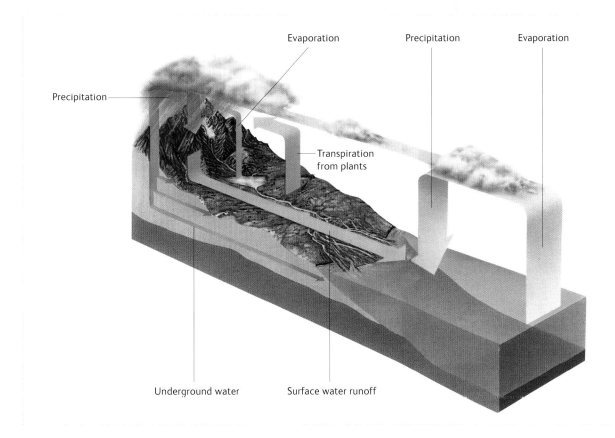

Precipitation

Evaporation

Precipitation

Evaporation

Transpiration from plants

Underground water

Surface water runoff

RIGHT A 'black smoker'; seawater that has circulated through the ocean crust, been heated and enriched in metals, forms mineral deposits when it mixes with cold, oxygen-rich water on the seafloor.

BELOW. Evaporation of seawater in a lagoon with restricted flow from the open ocean causes salinity to rise and eventually results in evaporite minerals to be deposited.

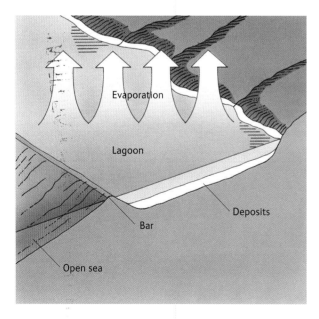

marine organisms. The chemistry of seawater is also profoundly influenced by active recycling of material between ocean and crust. Ocean water is constantly passing through underwater volcanoes at the ridges of spreading plate margins. This water interacts with the hot volcanic rock and extracts chemicals into seawater, and is also responsible for precipitating metal-bearing minerals on the seabed.

Crystallisation that results from evaporation is the most obvious, and easily detectable of these processes. Just as a crust of minerals leached from the soil of a watered pot plant is left behind as the water evaporates, minerals are left behind by evaporating seawater. Evaporite deposits can form where bodies of seawater become enclosed or temporarily isolated. When seawater is concentrated to about 50% of its original volume by evaporation, minerals begin to crystallise out of solution in reverse order of the solubilities – the least soluble crystallising out first. The chief marine evaporites are gypsum (calcium sulphate hydrate) and halite (sodium chloride) or common salt.

Today, the annual rate of evaporation in the Mediterranean Sea is over 4,000 km^3 (4,000,000,000 litres) of water, of which only 10% is replenished by rainwater and rivers. The rest still comes from the Atlantic, and if this source were to be cut off, the Mediterranean would dry out. This did happen, 6 million years ago. The Mediterranean Sea was closed at its western end, and the trapped water – some

4 million km³ (4,000,000,000,000 litres) – gradually evaporated, leaving behind a temporary desert encrusted with a hard layer of evaporites. Five-and-a-half million years ago, the link to the Atlantic was re-established and the sea returned. Gradually new marine sediments buried the evaporite layers hundreds of metres below the seabed.

Seabed soundings of these layers show remarkable structures called salt domes, a few kilometres in diameter and many hundreds of metres thick, penetrating into the overlying sediment beds. Under high pressure salt slowly begins to flow and exhibits 'plastic' behaviour. As it is more buoyant than the overlying sediments, it will flow upwards wherever it finds a weakness, eventually forcing a way through overlying rock strata. Salt domes are economically important because they can create traps for oil and gas to accumulate. By disturbing the rock strata, they block the flow of these deposits through more porous rock layers.

Seawater is only 3.5% dissolved material, which means that evaporation of even large, isolated seawater masses would not leave more than a thin crust of deposit. But some ancient evaporite deposits are hundreds of metres thick, so the bodies of water from which they crystallised must have been replenished at regular intervals with new saltwater. Thick ancient salt deposits are mined around the world for instance in countries around the southern North Sea in Europe and in the southern United States. Today, the environments in which evaporates form are the sabkhas, coastal flats in the Middle East periodically replenished by the sea, and land-locked salt lakes or playas such as those in California's Death Valley, occasionally fed by rainstorms and flash floods (see p.66).

LIFE INTO ROCKS

Not all rocks at the Earth's surface are formed from the debris of older rocks. Some, such as evaporites, are chemical deposits. And others are made from the debris of long-dead organisms, mostly of marine origin. In the oceans, billions of microscopic organisms, both plants and animals, depend on the seawater in which they float to supply them with the nutrients and dissolved materials they need. They remove dissolved calcium carbonate from the water to build their bodies, and when they die their remains sink to the ocean floor, locking up the minerals they contain in sediments called oozes. Shellfish and other larger invertebrates build their shells and other hard parts in the same way and their remains also accumulate in sea floor sediments.

Two fates can befall the hard parts of marine organisms. They may become mixed with sand along a shore for instance and, if buried and cemented, incorporated in a sandstone. However, if they accumulate where little or no sediment is entering the sea from the land they can be buried to form limestones. Limestones vary in composition and purity depending on how, where and from which organisms' remains they were formed. Many are made up of fine-grained calcium carbonate derived from marine organisms as well as recognisable shell debris. A proportion of

BELOW Section through a core of rock salt (halite) from the southern North Sea.

RIGHT Shelly limestone about 420 million years old – a fine example of a marine limestone deposit, showing fossilised remains of sea creatures, from Dudley, West Midlands, UK.

ABOVE Coccolithophorid, a type of marine phytoplankton. When deposited on the seafloor the sphere you see often falls apart into the ring-shaped coccoliths that define its shape. Coccoliths are made of calcium carbonate and make up the chalk. Size about 15 microns in diameter.

sediment eroded from the land may be present or the limestone may be over 80% pure calcium carbonate, as is chalk.

Stromatolites are built from mats of cyanobacteria (blue-green algae) or other microscopic algae, which grow as a gluey film over rocks in tidal flats. They build up layer upon layer, each new growth trapping a fine layer of carbonate. Ancient stromatolites are important evidence of the earliest life forms on Earth.

While living organisms store carbon temporarily in their bodies, their accumulated and lithified remains form a longer-term store. Deposits that are made from the remains of living things, rather than the fragments of other rocks, are important components of the Earth's surface, and important 'stores' of carbon. When most of us think about carbon stores we probably think of coal and other fossil fuels, oil and gas. Although economically the most important examples, these constitute only a very small percentage of deposits on Earth that contain carbon. Most is stored as calcium carbonate in limestones such as chalk, ancient coral reefs and the vast oozes of sediment on the ocean floor.

Many ancient limestones have been converted to the mineral dolomite, which is a carbonate of magnesium and calcium. It is thought that the limestone has undergone secondary chemical change as magnesium-rich waters percolated through the existing limestone (calcium carbonate), with the magnesium gradually replacing some of the calcium.

LIMESTONE LANDSCAPES

Limestone can form tough rocks, but will dissolve away completely under attack by acidic water. So in limestone regions, very distinctive geomorphological features and landscapes are created.

Many limestones are cut by a regular pattern of fractures or joints through which water can readily flow. As a result there is typically little surface water in limestone hills as most of it sinks down swallow holes and trickles through joints, to re-appear as a stream somewhere at the base of the rock. On the surface, water erosion may enlarge the joints into a pattern of clefts and blocks, called a limestone pavement.

Beneath the ground, the action of water continues. Just as streams erode valleys and create floodplains, underground water etches out the rock to form great underground channels and caves. A cave is fundamentally an erosional feature, but we tend to recall the precipitated deposits that form in them. Where percolating

BELOW Typical features of a limestone landscape, formed as the rock is dissolved by slightly acidic water.

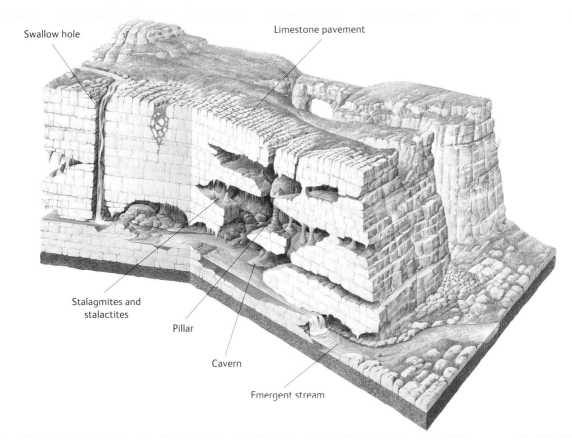

Swallow hole

Limestone pavement

Stalagmites and stalactites

Pillar

Cavern

Emergent stream

RIGHT The Pinnacles of sharp-edged limestone rise 30 m (100 ft) through the forest canopy of the mountain of Gunung Mulu in northern Sarawak.

water re-deposits minerals, spectacular structures can develop. An increase in temperature or decrease in pressure will cause loss of water by evaporation, and result in precipitation of minerals. Excess calcium carbonate is deposited on walls, ceilings and floors of the caverns creating dripstone, stalactites and stalagmites.

Karst scenery is a term applied to a characteristic weathered limestone landscape, where many of the limestone deposits are massive and consist of relatively pure calcium carbonate. (Karst is the German form of a Slovene word for 'bare stony ground'.) The scenery is characterised by a lack of surface drainage, a thin soil, and depressions where surface water sinks into the ground along massive joints, widened by dissolution, and runs underground to form caverns. As these caverns enlarge

in time they may eventually collapse creating a hole at the surface. One notable example occurred in Florida, in May 1981, when a three-bedroomed house, with half its swimming pool and six cars from a neighbouring parking lot, collapsed into an underground cavern. The hole created eventually broadened to 200 m (656 ft) wide, and 50 m (164 ft) deep.

'Pinnacle' karst, such as the extraordinary sharp-edged topography of Sarawak, is formed in humid tropics by chemical dissolution of metamorphic limestone, with high rainfall, high rates of dissolutions and high levels of biological activity. The 'tower' karst of Guilin, China, has broad alluvial valley floors and gorges between high towers of limestone undercut by the river water.

ENDANGERED SPECIES?

Limestone pavement is a horizontal limestone surface, naturally occurring, with deep grooves called 'grikes' between flat-topped areas called 'clints', where the rock has dissolved away along the lines of joints in the bed. Quite how this pattern formed is a matter of debate. Scouring glaciers have exposed the bare rock to acid rainwater attack, although the process of dissolution may have already begun while soil still covered the rock. But limestone pavements require further explanation, as many other limestone areas do not show this distinctive topography even though they too were stripped by glacial erosion. The exact way in which massive limestones dissolved is more likely to be related to the cementation structure of the particular limestone.

Limestone pavements take thousands of years to form, and are a unique natural habitat. (The most famous examples occur in the European Alps, Ireland, northern England and Sweden). Unfortunately, it takes little time to cut up and remove the rock, and in Britain and Ireland it has been in great demand for garden rockeries. Few areas of limestone pavement have been left undamaged, so that the remaining areas are now protected. The strange beauty of these environments also poses a threat to their long-term future – as millions of tourists clamber over them every year, eroding their surfaces further.

ABOVE Malham Cove, Yorkshire, UK –
a classic limestone pavement.

LEFT Stalactites and stalagmites in the Carlsbad Caverns, New Mexico, USA.

THE CARBON CYCLE

Atmospheric carbon dioxide is ultimately the source of all carbon in living things, at the start of the chain that leads from plant photosynthesis to the formation of carbon-containing sedimentary rocks. Formation of the rocks removes carbon dioxide from surface reservoirs, to be immobilised in the Earth's crust. Human activities release it much faster than natural processes.

On a global and geological scale, the constituents of the Earth's surface are not fixed, but are continually exchanged between atmosphere, surface and interior. One of the most important of such cycles, for us and all other life on Earth, is that of the element carbon. The carbon cycle is really a summary description of how carbon travels through and between living and non-living realms – between air, water, life and rock, and between the surface and interior of the Earth. At times, carbon

COAL

The coal series (**FROM TOP**) plant material, peat, lignite, bituminous coal and anthracite. As the coal-forming material is buried and compressed it undergoes physical and chemical changes. Water and volatile gases escape, leaving the deposit richer in carbon. Deeper and longer burial concentrates the carbon.

In the presence of oxygen, microbes breakdown detritus from land plants and release carbon dioxide into the atmosphere. But in a low-oxygen environment, such as a swamp or stagnant lake, the process of decay is lessened with first peat and, then after deep burial, coals being produced.

Coal deposits are sedimentary rocks and often occur in a sequence, interleaved with beds of shale and sandstone, which records a rhythmic pattern of past change. Deposition of sand then mud by rivers led to silting-up and formation of a swampy area in which plants could begin to grow. The plant material rotted down to form peat, was buried by incoming mud and compressed, eventually to form coal. Subsidence allowed rivers to cover the swamp area again and the cycle repeated to form the series.

The formation of coals, and other sources of energy such as oil or natural gas – 'fossil fuels' – have trapped the energy of growing living things. In doing so they have concentrated the carbon in the carbon dioxide that occurs in the atmosphere. By burning fossil fuels, humans release this stored carbon back into the atmosphere as carbon dioxide. Before industrialization carbon dioxide comprised about 0.03% (280 parts per million) of the atmosphere. Today, it makes up nearer 0.04% (380 parts per million) of the atmosphere.

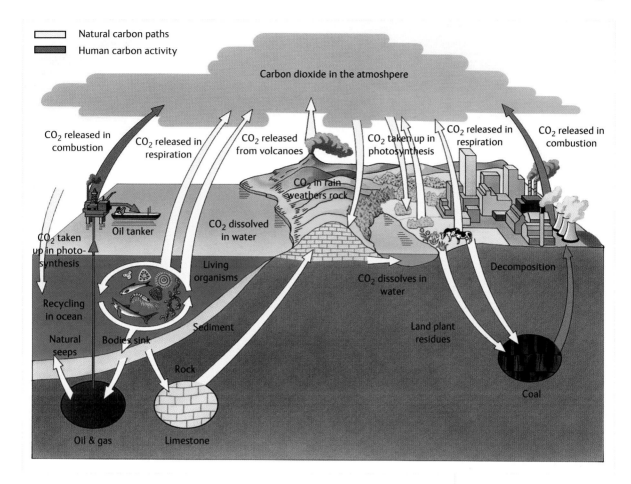

Natural carbon paths
Human carbon activity

Carbon dioxide in the atmoshpere

CO_2 released in combustion

CO_2 released in respiration

CO_2 released from volcanoes

CO_2 taken up in photosynthesis

CO_2 released in respiration

CO_2 released in combustion

CO_2 in rain weathers rock

CO_2 dissolved in water

Oil tanker

CO_2 taken up in photo-synthesis

Recycling in ocean

Living organisms

CO_2 dissolves in water

Decomposition

Natural seeps

Bodies sink

Sediment

Land plant residues

Rock

Coal

Oil & gas

Limestone

ABOVE **The carbon cycle.**

is captured and stored in a carbon sink such as a soil, limestone or coal deposit, then it is released again. Sometimes it may remain locked up for tens or hundreds of millions of years, sometimes much less. By excavating and burning fossil fuels, we short-circuit the much longer natural cycle to release carbon dioxide back into the atmosphere. This is contributing to current global warming that will have as yet unpredictable effects on the planet.

ROCKS AND THE LANDSCAPE

C HANGES AT THE SURFACE OF THE EARTH, and the processes and products of change, occur at all scales, from the microscopic to the global. But for us, as human beings, the most important scale is our own – the landscapes we live in and survive on, why they are the way they are, and how they are changing. This chapter looks at some typical landscapes, and examines what clues they can give to past environments.

- What types of rocks are present and how did they form?
- What are the dominant erosional landforms and how were they produced?
- What are the processes active today, and what changes will they bring about?

History imposes layer upon layer of change, and in time rocks may be deformed and heated. This masks earlier changes and imposes new patterns. As a result, some of

OPPOSITE Mountains in the Sierra Nevada, California built by plate tectonic processes and sculpted by the forces of erosion.

BELOW The Tibetan Plateau is one of the fastest-eroding places on Earth today.

525 million
years ago

340 million
years ago

248 million
years ago

95 million
years ago

ABOVE These maps show how the continents have moved across the Earth's surface over the last 525 million years, at times colliding to form bigger landmasses then later breaking apart into smaller landmasses.

the answers to the above questions may be difficult to disentangle, and advanced geological detective work may be required.

In recent geological time over much of both northern and southern hemispheres, ice has been the dominant sculptor, altering earlier landscapes dramatically. Only a short while ago, ice covered not only the polar regions but large tracts of other areas, scouring them with ice sheets and glaciers. As water was locked up in ice, land once below sea level was exposed and subjected to the processes of subaerial erosion. The ice sheets have gone through long-term cycles of advance and retreat, and we are now in what was thought to be another 'interglacial' period, which would lead to other ice ages. But, alternatively, there is now the possibility that we might be entering a warmer period, brought about by global warming of our own making. And this will bring different surface processes to bear on the landscape.

Going further back in time over 300 million years, America was joined to what is now Scotland. This ancient continent lay in quite a different part of the globe, subject to a different climate. Scotland was south of the equator, but England was separated from it by an ocean. As these continental masses both slowly drifted north, they also slowly collided, forming a spectacular belt of mountains in the process. By then, the continent had moved sufficiently far north to lie within an arid climate belt, and desert conditions prevailed. Hot desert winds eroded the new mountains, breaking them down into sand that was deposited between the mountain ranges to form sandstones which are being eroded again today.

Later on, as the continent drifted further northwards into tropical climes much of the land surface was underwater. When the seas were at their most extensive, vast quantities of carbonate sediment accumulated and compacted to form chalk. Later, this was uplifted in further tectonic movement, exposing it to erosion before new sediments were unconformably deposited on it. So the rocks of the land formed and reformed, creating new sediments and new rocks out of old, in a never-ending cycle which continues today.

Much of the east coast of England is now being worn away. Subjected to the battering of the waves, the coastline is changing shape - landslides are common and the coast is retreating some 2 m (6½ ft) every year. Since the Domesday Book recorded the settlements of England in 1086, 26 coastal villages have disappeared before the waves. The east coast's erosion is hastened because it is also slowly sinking, as the Earth's crust in the area re-equilibrates itself after being weighed down by the last great ice sheets. (The ice weighed more heavily on the west part of the landmass and tipped the east upwards.)

Throughout the world, landscapes record geological change. In many areas of the world not now deserts, you can find the marks of ancient deserts in the rocks. The cross-bedded sandstone cliffs in Utah were formed 280 million years ago, when the land lay near the equator, within a desert climatic belt. Hot winds eroded the older mountains and the evidence of their action can be seen in the rocks. Individual sand grains look like millet seeds, rounded and polished by wind abrasion.

LEFT Cliffs of sandstone deposited in a desert being re-eroded in the hot, dry climate, Utah, USA.

A DESERT LANDSCAPE

BELOW Desert landscape – sparse vegetation survives between the exposed rocks.

The high Mojave Desert of California, and Death Valley lying to its northeast, are lands of extremes – some of the hottest, driest places in the northern hemisphere. Temperatures can average 49°C (120°F) in summer. As moisture-laden winds from the Pacific Ocean drop their rain when they rise over mountains near the coastline, the rainfall in the desert is low. There is no soil and little vegetation.

Here one can find an extraordinary range of desert landscapes: sand dunes and flat, gravelly pediments; rugged honey-coloured mountain slopes; deeply eroded canyons; and wide valley plains with stagnant marshes and salt flats. These landforms, so obviously shaped by surface processes, owe their existence to major tectonic activities.

Death Valley itself has been called a 'lesson in geology'. The valley was formed some 30–35 million years ago, when complex plate movements stretched and pulled the crust apart in this region of North America. (These same movements formed the notorious San Andreas Fault.) Where the crust was pulled apart, blocks between parallel sets of faults dropped down – in a process called block faulting – to form deep valleys. At its lowest point, at Badwater Basin, Death Valley has the lowest elevation in North America, some 86 m (282 ft) below sea level.

BELOW Badwater Basin, USA – the lowest place on the North American continent.

Eroded sediments filled the sinking valley, and intermittent volcanic action triggered by the crustal movements added further sediment. In places, the sediment is so deep that bedrock lies almost 3 km (1¾ miles) below the surface. The valley has also filled periodically with lakes which then evaporated to leave behind thick layers of salts, which were once mined.

Despite the very low rainfall, water still dominates the shaping of this landscape. Around 200 million years ago seas covered the land, depositing deep layers of sediments that are now thrust up into hills and mountains. Today, water cuts deep into these layers, scouring the hill-slopes of any loose soil and sand and carrying this material down steep-sided wadis to drop it in wide alluvial fans. And the continuing periodic flooding, then evaporation of water, feeds the saltpans which are so characteristic of this landscape.

LEFT Deep erosion by water creates networks of gullies – typical 'badland' features.

BELOW Lulworth Cove, UK. The sea has broken through hard outer cliffs and eroded the soft sediments behind to form the cove.

A COASTAL LANDSCAPE

The coast of East Devon and Dorset in England is a natural World Heritage Site along which rocks recording 185 million years of Earth's history are exposed in 153 km (95 miles) of dramatic cliffs, sea arches and stacks. The wonderful rock sections and beauty of this coast are maintained by the power of wave erosion and are the products of a retreating coastline.

Along the coast around Lyme Regis and Charmouth in Dorset, the cliffs are made up of alternating limestones and mudrocks in a succession called the Lias. This rock formed as soft sediments at the bottom of a shallow sea, 196–203 million years ago. We know that the sea was then teeming with now-extinct creatures such as sea-crocodiles, icthyosaurs, plesiosaurs and ammonites, because their fossilised remains are found in the Lias rocks. As you walk along the shore, especially after heavy rain, you can often see what looks like small mudslides where parts of the Lias have slid on water-saturated beds of mudrock. This continuous natural erosion provides an ongoing supply of fossil specimens for the thousands of visitors who visit the area every year often discovering the wonders of geology there for the first time.

Further along the same coast, Lulworth Cove is a small, almost circular bay with a narrow opening to the sequence of hard and soft layers of sedimentary rocks. The oldest rock is a 150-million-year-old, hard, shelly limestone. Younger beds lain immediately on top of it are of softer material and are more vulnerable to erosion.

BELOW Charmouth, UK – the soft Lias cliffs are vulnerable to wave erosion and human attack.

These in turn were covered by relatively resistant chalk. All the beds have been dramatically tilted to near vertical, and in part crumpled into a large S-shaped fold, so that the older, hard limestones form the cliffs. In one place the sea has eroded through these limestones and formed the narrow entrance to the cove. Further erosion by the sea has eaten away the soft sediment that, because of tilting, now lies behind the limestone forming the bay itself. The chalk has resisted rapid erosion and forms a high cliff at the back of the cove.

AN ICE-FORMED LANDSCAPE

Some of the most beautiful ice-formed landscapes in Britain are to be found in Wales, Northern Ireland and Scotland, shaped by great ice sheets riding over old mountains on numerous occasions over the past two million years. Ice has far greater powers of erosion than any river: it can fill entire valleys and scoop out rock from valley walls, leaving deep U-shaped valleys such as the dramatic Glencoe in Scotland. This is a landscape still bearing fresh wounds of erosion.

BELOW Eroded granite peaks rising along the sides of Glencoe, Scottish Highlands.

Glencoe is bounded by almost sheer rock faces of ancient granite, with cliffs up to 1,000 m (3280 ft) high at its pass. Here, the advancing ice breached the mountains, possibly taking advantage of an ancient river valley. The ice cut down an oversteepened valley into weaker metamorphic rocks which lay between huge granitic intrusions. Some of the rocks at the base and the walls of the valley have been scarred and polished by the abrasive action of the ice. The valley has been left littered with the glacial debris carried and dumped by the melting ice. On nearby Rannoch Moor lie some of Britain's most spectacular erratics, originally plucked from the ground in Scandinavia and transported hundreds of kilometres in ice. To the west, Loch Etive is a classic glacial fjord, a submerged glacial valley as deep as the fjords of Norway. Its straight path demonstrates the bulldozing power of the ice that formed it.

RIGHT Rannoch Moor, Scotland, scoured by ice and now scattered with erratics originally eroded from Scandinavia.

LEFT Glencoe, Scotland – the present-day river seems dwarfed in comparison with the deep, wide trough carved by bulldozing ice.

RIGHT The Whitsunday Islands, Queensland, Australia are in fact the peaks of drowned mountains.

BELOW The outer reef of the Great Barrier Reef is unprotected from the powerful waves of the Pacific Ocean, and tropical cyclones can damage it greatly.

OPPOSITE In the calm waters in the lee of the outer barrier of the Great Barrier Reef, corals grow in profusion. The main threats to survival are probably tourism and the effects of climate change.

A LIVING LANDSCAPE

The Great Barrier Reef is the largest coral reef on Earth, a living structure covering some 230,000 km² (142,915 miles²). Rather than one reef, it is in fact an intricate series of reefs, stretching more than 2,000 km (1243 miles) along the eastern coast of Australia and separated from the mainland by a shallow lagoon.

Some of the coral is many millions of years old, but most is much younger. The story of this younger coral is intimately linked with the last Ice Ages, when huge volumes of water were withdrawn from the seas into vast ice sheets. This exposed a wide, shallow continental shelf, thick with sediments, to the east of the Australian mainland. The newly dry land was eroded into hills and valleys, with a range running along the outer rim. Many of the islands on the reef, such as the Whitsunday Group,

are the peaks of drowned ridges along this range, now fringed with coral. By 12,000 years ago, the sea was rising again, flooding the land. Corals began to grow in the shallow water, keeping pace with the rising sea levels and forming the elaborate series of reefs and lagoons we see today. The reef reached its current level about 6,000–7,000 years ago. Some parts are now well below the growing depth of living coral, extending down some 60 m (197 ft).

The growth of coral reefs is strongly influenced by the water's warmth, salinity and transparency to sunlight. Reef corals can only grow in water temperatures higher than 17.5°C (63½ °F), and usually no deeper than to about 50 m (164 ft) below the surface. They cannot tolerate low salinities or cloudy waters, and it is the influx of sediment-laden freshwater that determines the northerly limit of the reef.

STUDYING CLIMATE CHANGE

I N STUDYING CLIMATE CHANGE one thing is obvious: we are dealing with a highly complex surface system whose nature is inherently changing. A myriad of interlinking factors (apart from the generation of an enhanced greenhouse effect) affect the outcome, and all the surface systems – air, water, rock and life – are intimately involved.

Many feedback mechanisms exist which help to regulate climate: some negative, damping down the change; others positive, accelerating change. The difficulty lies in predicting which will come into operation, and how these mechanisms will interact.

When temperatures rise, for example, the evaporation of water from sea and land increases. The result is an increase in cloud cover, but this could have one of two opposite effects on climate. Cloud cover reflects solar radiation back into space. This is called the albedo effect. (Albedo is the reflectivity of the Earth's surface. Snowfields and glaciers have high albedos and reflect 80–90% of sunlight. City buildings and

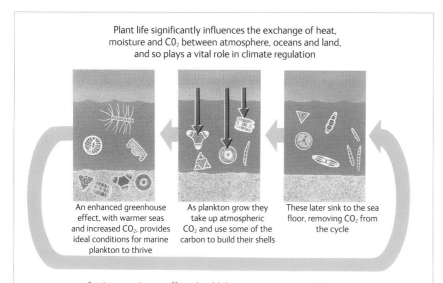

Plant life significantly influences the exchange of heat, moisture and CO_2 between atmosphere, oceans and land, and so plays a vital role in climate regulation

An enhanced greenhouse effect, with warmer seas and increased CO_2, provides ideal conditions for marine plankton to thrive

As plankton grow they take up atmospheric CO_2 and use some of the carbon to build their shells

These later sink to the sea floor, removing CO_2 from the cycle

So the greenhouse effect should decrease, in a negative feedback loop

OPPOSITE Increasing energy use throughout the world is leading to ever-increasing carbon dioxide emissions in the atmosphere.

LEFT Negative feedback loop: enhanced CO_2 in the atmosphere encourages planktonic growth. In growing, the plankton may eventually absorb the extra CO_2 to restore the balance but major impacts, such as ocean acidification, may occur before this has time to happen.

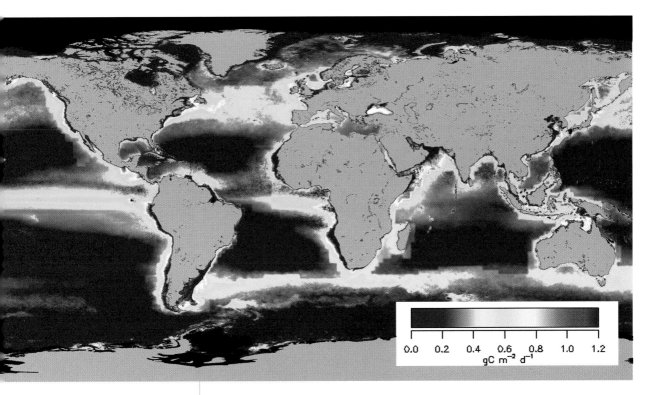

0.0   0.2   0.4   0.6   0.8   1.0   1.2
gC m^{-2} d^{-1}

ABOVE Computer-enhanced satellite image showing biological productivity of plankton in the oceans (as you move towards the red, productivity increases). The growth of such marine plankton takes up carbon dioxide from the atmosphere.

dark pavements have albedos of only 10–20%.) Increased cloud cover could lead to a fall in surface temperatures, as more solar energy is reflected, in a *negative* feedback mechanism. But increasing cloud cover might also have a *positive* feedback effect, because water vapour is the most important greenhouse gas of all, and an increase could lead to greater retention of heat inside the Earth's atmosphere.

A potential positive feedback loop also exists to enhance the albedo effect when temperatures *fall*. As temperature decreases, ice cover increases, and less heat is absorbed. Surface temperature therefore continues to fall.

As research continues, more and more subtle and complex interactions are discovered which must also be considered to try to understand, and perhaps to predict, climate change. Scientists build highly complex computer models called 'global circulation models', which allow them to undertake simulated experiments on the Earth's response to changes such as the increased burning of fossil fuels. An imaginary atmosphere is modelled, based on data on the known physics of the atmosphere, known patterns of climate change and current understanding of the complex interactions between air, sea, ice and land. Billions of calculations are involved in modelling the climate system and how it might evolve. The results will depend on the assumptions upon which a particular run of the model are based. Different sets of assumptions are used to generate different 'predictions' that can be compared to give the range of possible likely climate changes. Data relating to past climate changes, whether details of changes in the recent past or inferences from the rock record, can be used to evaluate whether a prediction is reasonable or not.

SOURCES OF EVIDENCE

Studying the evidence of past climates is a vital part of learning about the Earth's climate system, and how and why change came about. Many kinds of evidence can be used. For example, pollen and other plant material provide useful indicators of past climates, while fossil corals show growth patterns that record changes in climate and sea level. More subtly, tiny planktonic fossils also provide a record of past climates. For example, *Globigerina pachyderma* is like a fossil thermometer: in cold water its shell grows by coiling to the left, in warmer waters it coils to the right.

Ice cores taken from the Antarctic and Greenland ice caps can also provide a record of past temperatures, as the ice holds 'prehistoric air' trapped in tiny bubbles, which can be extracted and its oxygen analysed. Oxygen exists as two isotopes, O^{16} and O^{18}, and the balance between the amount of these two isotopes dissolved in water varies with temperature: the colder it is, the greater the proportion of O^{18} in the water, the less in air. The air thus has an isotope 'signature' which can provide useful evidence of prehistoric temperatures. A 3.27 km (11 miles) long core from Antarctica represents 800,000 years of snow accumulation and eight glacial cycles. Interestingly, evidence from ice cores suggests that the climate shifted suddenly rather than gradually between cold snaps and warm spells. This presents a major challenge for the near future; we may be faced with the consequences of our actions more suddenly that we have been expecting.

ABOVE A glaciologist in Antarctica extracts an ice core from the apparatus which has drilled it from deep within the Antarctic ice sheet. The glaciologist, a member of the British Antarctic Survey (BAS), is dressed in protective garments in order to minimise contamination of the ice core.

COMPLEXITY AND CHANGE

In addition to changes in the composition of the atmosphere, other natural factors can affect climate. Changes in solar radiation, tectonic events, slow variations in the Earth's orbit around the sun, and catastrophic events such as volcanic eruptions and occasional large meteorite impacts, may all be important.

Astronomers have long appreciated that the sun's output of heat and light is not constant. Sunspot activity provides a measure of this variation and comparisons have been made between sunspot frequency and climate change to see if the two are linked. The nature of the connection (if any) is still under debate, but the onset of the Little Ice Age, a period of extreme cold in 17th and 18th century Europe, has been correlated with contemporary Chinese records of increased sunspot activity.

Changes in the Earth's orbit round the sun have also been suggested as an explanation for the rapid advances and retreats (reflecting temperature changes) of the last ice age. Three types of orbital changes occur:

ABOVE Frost Fair, London, 1613; during the Little Ice Age, the River Thames in London regularly froze.

1 The shape of the Earth's orbit around the sun changes in a cycle of about 100,000 years.

2 The Earth is tilted on its axis, and this tilt oscillates by 1.5˝ in a cycle of about 41,000 years.

3 The Earth's axis, which now points directly towards the North Star, wobbles like the axis of a spinning top as it circles the sun, completing a full cycle every 23,000 years.

In the early 20th century, a Yugoslav scientist called Milutin Milankovitch calculated the combined effects of these cycles and showed how their effect on climate coincided with the known timing of ice age advances and retreats.

The cause of the onset of the ice ages is still a mystery, however, although there are many theories. One theory proposes that the raising of the Himalayas and the Tibetan Plateau, with the collision of the Indian and Asian landmasses, was the trigger.

Tectonic movements are important in influencing global climate. The distribution of plates and their continents over the globe, and the relative position of landmasses and oceans, determine the global circulation of wind and water. Evidence from the geological record suggests that the great ice ages have occurred

STUDYING CLIMATE CHANGE **99**

during periods when the continents were clustered near to the poles. For example, during the Permian, all the continents were gathered near the South Pole into one giant supercontinent called Pangea, and glaciation was extensive at this time. In the more distant past, around 650 and 725 million years ago and possibly during earlier episodes, much of the Earth may have been covered by ice. Such times of 'Snowball Earth' have been proposed because rocks formed of glacial sediments are found in areas that at the time would have been in the tropics. This idea is not accepted by all but if correct indicates that the global mean temperature would have been about -50°C (-58°F), because most of the sun's radiation would have been reflected back to space by the icy surface, the average temperature at the equator would have been about -20°C (-4°F), similar to that of Antarctica today.

Volcanic eruptions of material from the Earth's interior can also influence climate. Major eruptions can eject large quantities of ash and fine particles high into the atmosphere for them to be carried around the world. These can persist in the atmosphere for several years and reduce the amount of solar energy reaching the Earth's surface. Large amounts of gas can also be released by eruptions. The Krakatoa eruption in 1883, for instance, ejected large volumes of sulphur dioxide as well as 25 km³ (25,000,000 litre³) of rock and ash. This gas would have been transported throughout the stratosphere and reacted to increase the sulphurous acid content of high-level clouds. This change in the composition of the clouds would have increased their albedo and so, along with the ash, reduced the solar radiation reaching the Earth's surface. Overall the erupted material is believed to explain a drop of just over 1°C (34°F) in the average global temperature over the year following the eruption. It took 5 years for temperatures to stabilise at their pre-eruption levels.

Periods of prolonged eruptions in the geological past may have had more profound effects on climate. Between 60 and 68 million years ago a vast volume of lavas was erupted in India covering an estimated area of 1,500,000 km² (932,056 miles²) (approximately half the size of present-day India) to a thickness of more than 2 km (1 mile). These eruptions were not continuous but it has been suggested that they released large volumes of greenhouse gases and resulted in significant changes in climate. One interpretation is that average global temperature increased by 8°C (46°F) over a period of half million years and this change contributed to the extinction of the dinosaurs and a wide range of other species.

Geological events other than volcanic eruptions may play their part in climate change. Sudden release of the greenhouse gas methane may have occurred at times in the past. Methane can form solid chemical structures called clathrates with water at low temperatures and moderate pressures. The conditions are right for methane clathrate formation under the seafloor along the outer continental shelf and large volumes of methane are stored there today. At times in the past such clathrates may have become unstable because of warming or reduced pressure with the result that the methane was suddenly, in geological terms, released into the atmosphere.

Shifts in the Earth's position relative to the sun, changes in the output of solar energy, motions of plates and the resulting movement and collision of the

INDIA HITS ASIA – AN EVENT OF GLOBAL IMPACT

Of all the changes that have occurred on the Earth's surface over the past 50 million years, perhaps the most dramatic has been the collision of the Indian continent with Asia, and the formation of the Himalayas and Tibetan Plateau. This extraordinary event was itself only part of a constant pattern of change, and it wrought fundamental changes in all the global systems, and, in turn, is shaped by them. The Himalayas are the highest mountains on Earth, yet one day they will be worn away to nothing under the destructive power of surface processes.

As the Himalayas rose, they changed patterns of airflow between land and sea. The monsoons intensified as the growth of a high landmass changed the path of the rain-bearing air. The Gobi Desert formed in an area of once lush ground as the rising Himalayas created a rain-shadow. The intensification of rain over a high relief affected the pattern of weathering dramatically. The Himalayas and Tibetan Plateau are today the most rapidly eroding landscapes in the world; already 6 km³ (6,000,000 litres³) have been removed from the land. It has been proposed that this intense weathering drew down carbon dioxide out of the atmosphere so rapidly that it triggered the final dramatic cooling event that led to the ice ages.

This change was only one element in a great global rearrangement of oceans and land which also led to the creation of the Atlantic

TOP Himalayas: the world's highest mountains.
ABOVE India drifts north to collide with Asia.

Ocean and the dramatic realignment of global ocean currents. This must have caused major shifts in the distribution of heat and has itself been proposed as an alternative trigger for the start of the last ice ages. Which theory – or combination of theories – ultimately proves to be correct, will be known only after further scientific investigation and debate.

continents, and outbursts of volcanic activity may have all influenced global climate. These operate, however, on different timescales. Plate motions cause changes over million of years. Prolonged volcanic activity may bring about marked changes over hundreds of thousands of years or longer, while single eruptions influence climate over a matter of years. In the background changes in the Earth's orbit around the sun and the Earth's rotation subtly affect change over tens of thousands to a hundreds of thousand of years.

CARBON DIOXIDE AND CLIMATE CHANGE

Humans have been adding greenhouse gases to the atmosphere at an increasing rate since the Industrial Revolution as a result of burning fossil fuels and agricultural activities such as deforestation. Carbon dioxide is the most important of these additions but humans have also increased the concentrations of two other important greenhouse gases - methane and nitrous oxide. Overall human emissions of greenhouse gases increased by 70% between 1970 and 2004.

These were some of the findings of the fourth assessment report of the International Panel on Climate Change (IPCC) published in 2007. Scientists and other experts from over 130 countries contributed to this report over the previous 6 years. The IPCC report concluded that "there is very high degree of confidence that the net effect of human activities since 1750 has been one of warming" and "it is likely that there has been significant warming (as a result of human activities) over the past 50 years averaged over each continent (except Antarctica)".

Some people still question this link between human activities and global warming saying that the Earth's climate has always varied or that recent changes might be explained by sunspot activity on the sun.

The Earth's climate has indeed varied over geological time, experiencing the ice ages in the last million years or so and events like the 'Medieval Warm Period' in historical times. Such changes result from a variety of factors which interact in a complex fashion. The ice ages can be related to the path of Earth's orbit around the sun. Volcanic eruptions puts dust into the atmosphere that can remain there for some years, or release carbon dioxide. Increased sunspot activity indicates greater solar energy emission. But when all such factors are taken into account the increase of 0.74°C (33°F) in average global temperature that has occurred over the last century is larger than can be accounted for by such natural factors alone. For instance, increased solar energy emissions as indicated by sunspot activity may have made a contribution to global warming in the early 20th century but satellite data show steady, or possibly decreased, sunspot activity over the last 30 years.

As far as scientists can tell, the questions are not, is global warming occurring and is this linked to human emissions of carbon dioxide and other greenhouse gases, but what will be the degree of the warming, what effects will this have and how can we minimise future warming and its effects?

THE GREENHOUSE EFFECT

Carbon dioxide in the atmosphere is essential to life on Earth. Along with other 'greenhouse gases' – such as water vapour, methane, and nitrous oxide – it helps balance the Earth's heat budget. Without these gases, the planet would be extremely cold and as inhospitable

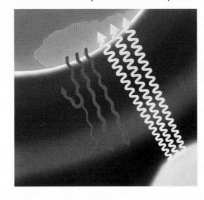

to life as the moon. Incoming solar radiation that is not reflected back into space by the upper layers of the atmosphere is absorbed by the rocks, waters and living organisms of the Earth's surface. In return, the warmed surface re-radiates energy out into space in the form of long-wave infra-red energy (heat). This energy is partially reabsorbed by the greenhouse gases in the air, keeping the Earth warmer than it would otherwise be by some 30°C (86°F). This is an entirely natural process, and makes life on Earth possible.

The greenhouse effect is so called because it has been likened to the trapping of warmth by the glass in a greenhouse. The glass acts like the atmosphere by letting in light, but keeping in heat. Global warming is thought to be happening because of human activities adding to the natural greenhouse effect.

The IPCC 2007 report concluded that "human influences have very likely contributed to sea level rise over the latter half of the 20th century, have likely contributed to changes in wind patterns affecting extra-tropical storm tracks and temperature patterns, and more likely than not increased risk of heat waves, area affected by droughts since 1970 and frequency of heavy precipitation events". The degree of global warming in the future and how the climate will respond cannot be predicted with certainty. Various likely warming scenarios and their affects have to be considered.

A 'middle-of-the-road' estimate by the IPCC suggests that the global average temperature may increase over the next century by 2 to 3°C (35 to 37°F). The possible impacts of such a change, identified by the IPCC, are varied and widespread. Water may become more available in the tropics and at high latitude but would become scarcer at mid latitudes increasing the risk of drought for millions of people. Farming yields would be likely to change, cereal productivity decreasing in low latitudes and some increasing in mid to high latitudes. On the coasts, where an ever increasing proportion of the world's population lives, the risk of floods and storms is likely to have increased, exacerbated by increased sea level resulting from melting ice and thermal expansion of seawater. Up to 30% of species would be at increasing risk of extinction, for instance with coral mortality being widespread.

Such impacts are considered possible on the basis of relatively well-known consequences and interactions. Some other effects are less easily judged. For instance uptake of carbon dioxide by the oceans since 1750 has led to the seawater becoming very slightly more acidic. Over the 21st century this slight increase in acidity is likely to increase further. The precise effects of such increases of acidity on

LEFT The UK and Eire – not two but a series of islands, if the world's ice melted.

marine organisms is as yet unrecorded. It is likely though to inhibit organisms which grow shells, for example corals, and so have an impact on ecosystems of which such organisms are important parts.

At the beginning of the 21st century, the link between increasing release of carbon dioxide and other greenhouse gases and global warming is demonstrated beyond reasonable doubt. How exactly global warming will influence climate change across the Earth is a matter for increased understanding and prediction. How changing climate will relate to changes in the atmosphere, hydrosphere and biosphere and interactions between them will be determined by ongoing observations and improved interpretations. How the pace of change can be slowed will be a matter for technological advances but above all else a matter for individual and political will.

GLOSSARY

ABRASION The mechanical wearing down of rocks, typically by rubbing or knocking against one another.

ALBEDO The ability of a surface to reflect the sun's heat and light.

ALLUVIAL FAN A mass of sediment deposited by a river at the base of a steep gradient.

ALLUVIUM A deposit, usually sand and gravel, transported by a river and deposited along its route, mostly on the floodplain.

ASTHENOSPHERE The mobile region of the mantle on which the Earth's plates move.

BED A layer of sediment. May be thin (see *lamina*) or span several metres.

BEDDING PLANE The surface of rock or sediment parallel to the plane of deposition of a bed.

BLACK SMOKER A vent on the sea floor, where heated fluid forming minerals escapes from the oceanic crust.

CARBON SINK Temporary store (although sometimes lasting millions of years) of carbon on or in the surface of the Earth, eg in *limestone*.

CEMENTATION The binding of loose sediment with a cement of material typically deposited round the grains from percolating water.

CHALK A fine-grained white *limestone* composed of microscopic *coccoliths*.

CLAY A sediment composed of the finest grade of weathered rock fragments (<0.002 mm diameter).

CLIMATE The average weather conditions in one area, usually classified by precipitation, temperature and vegetation.

COAL SERIES A series of beds, with seams of coal alternating with shale, mudstones and sandstones, typical of an estuarine swamp environment.

COCCOLITH Microscopic plates secreted by phytoplankton, composed of calcium carbonate.

COMPACTION The slow squashing together of grains as sediments accumulate.

CONVECTION CURRENT A current in air, water or molten rock, where heated material rises and flows outward from the heat source, and cooler material flows down and inwards to replace it.

CORIOLIS EFFECT The deflection of currents by the Earth's rotation.

CORALLINE ALGAE Lime-impregnated red algae.

CROSS-BEDDING A series of inclined bedding planes indicating migration of the bed with the direction of flow of an air or water current.

CRUST The outermost rocky 'skin' of the Earth.

DELTA A fan-like deposit of material where a river flows into a sea or lake.

DENUDATION Lowering of the land surface, the sum of the surface processes of *erosion* and transportation.

DESERT ROSE Evaporite structure found in desert sandstone, with the appearance of a series of rose petals.

DOLOMITE A limestone containing more than 15% magnesium.

DREIKANTER A three-sided *ventifact* weathered by sand abrasion.

DRILL-CORE A sample of rock, or sometimes ice, drilled out to examine its composition and structure.

DRIPSTONE An alternative, general name for *stalagmite*.

ERG Large expanse of sand dune deposits.

EROSION Wearing away and removal of rock (or land) either directly by agents of erosion or by *abrasion*.

ERRATIC A large boulder transported by ice some distance from its source to an area of different rock type.

EUSTASY Worldwide change in sea level, leading to either exposure of seabed or increased covering of land surfaces.

EVAPORITE Sediment left behind after evaporation of seawater or water from land-locked lakes (see *playa*, *sabkha*).

EXFOLIATION A process of weathering in which sheets of rock split off from the surface. Also known as 'onion skin' weathering.

FEEDBACK LOOP The continuous modification, adjustment or control of a process or system by its own results or effects. Negative feedback damps down change and positive feedback amplifies an effect.

FLOODPLAIN The broad expanse or flat land bordering a river, usually in its lower reaches, built up from *alluvium*.

FORAMINIFERAN (plural foraminifera) Geologically important marine protozoan, normally planktonic. Its carbonate 'shells' form thick layers of marine limestone.

FROST-SHATTERING The weakening and eventual fracture of rocks by repeated cycles of freeze–thaw.

GEOCHEMICAL Concerned with the chemistry of rocks and the patterns of chemical exchange between rocks, crust and other surface systems.

GLACIER A 'river' of slowly moving ice, flowing downhill under gravity.

GLOBAL MEAN SEA LEVEL The average sea level across the planet (averaged also across tides).

GLOBAL WARMING An increase in global mean temperature. Affected by many factors, the world's temperature has varied widely over its history. Today's global warming is thought to be due to a *greenhouse effect* enhanced by human activity.

GRADED BEDDING A pattern in a *bed*, where the particle sizes become finer towards the top of the bed.

GRAVITATIONAL POTENTIAL ENERGY Energy due to the gravitational pull of the Earth, tending to draw all things on the surface down to their lowest level.

GREENHOUSE EFFECT The trapping of heat radiated from the surface inside the Earth's atmosphere, thus keeping the planet warmer that it would be without an atmosphere.

GREENHOUSE GAS A gas in the atmosphere which reflects back heat radiating from the Earth's surface.

HANGING VALLEY A glacial landform, where the valley of a tributary glacier is cut by the main glacier, which cuts a deeper trough. When the ice retreats the tributary valley is left 'hanging', often forming a waterfall.

HYDROSPHERE The sum of water on the planet, from atmospheric water vapour to ice and underground reservoirs.

ICE AGE A cold period in the Earth's history when ice covered much more of the surface. The most recent ice age ended about 10,000 years ago.

IGNEOUS ROCK One of the three main types of rock (with *metamorphic* and *sedimentary*). Formed from once-molten *magma*, igneous rocks may have a crystalline or 'glassy' structure, eg granite, gabbro and basalt.

ISOSTASY The tendency of the Earth's crust to maintain an equilibrium position in the mantle, eg, slowly rising up after the removal of ice cover

KARST SCENERY Landscape showing typical features of limestone erosion, such as sink-holes and caverns.

KINETIC ENERGY Energy due to motion.

LAMINA (plural laminae) A very fine layer of sediment.

LATERITE Deep red tropical soil.

LEACHING The removal of minerals from soil by dissolution in water.

LIMESTONE Any sedimentary rock composed primarily of (calcium) carbonate. Most are of biological origin, although some have been chemically precipitated.

LIMESTONE PAVEMENT A flat area of exposed, eroded limestone, with deep grooves (grikes) cutting down between flat surfaces (clints).

LITHOSPHERE The outer 'skin' of the Earth, consisting of the crust and uppermost mantle.

LOESS A fine sediment derived typically from glacial outwash, and further redistributed by wind.

MAGMA Hot, molten rock from the Earth's *mantle* and lower crust, which crystallises to form *igneous rock*, either at the surface (as lava) or below it.

MANTLE The interior of the Earth below the surface crust and above the core. *Convection currents* in the mantle account for the movement of the *tectonic plates*.

MASS WASTING The mass-movement of material downslope under gravity, eg rockfalls, landslides, soil creep.

MEANDER Any bend that develops naturally in a river, gradually exaggerated over time.

MELTWATER Water that flows from the foot of a melting glacier (see *outwash*).

METAMORPHIC ROCK One of the three basic types of rock (with *igneous* and *sedimentary*), formed by the changing of other existing rocks under heat and pressure, below the surface. For example, limestone can be metamorphosed into marble.

MID-OCEANIC RIDGE A long ridge, like a long mountain range along the ocean floor, where two *tectonic plates* are separating. New sea floor is continually created as magma rises up from the Earth's interior.

MINERAL A naturally occurring crystalline single element or, more usually, compound, the basic constituent of *rocks*.

MORAINE The debris eroded and carried by a *glacier*, eventually deposited when the ice melts (see *till*).

MUD An unconsolidated sediment consisting of a mixture of *silt* and *clay*.

MUDROCK (or Mudstone) A rock formed from *mud*.

OOLITH A tiny sphere of calcite, formed from a 'seed' of shell fragment, rolled to and fro by waves in lime mud of shallow seas. (Many ooliths cemented together form oolitic limestone.)

OOZE A thick deposit on the deep ocean floor, largely composed of the remains of sea organisms such as plankton and shellfish.

OUTWASH The finer sand and gravel sediment from a glacier, carried onwards in the *meltwater* streams that run from its base.

PEDESTAL ROCK A wind-eroded rock found in arid environments, where the base is eroded faster than the top, often forming a mushroom shape.

PEDIMENT A gently inclined plain of eroded bedrock generally veneered with fluvial gravels.

PHYTOPLANKTON Aquatic, microscopic plant life-forms.

PLATE TECTONICS The movement of plates of the Earth's lithosphere, pulling apart at some margins, colliding, submerging and crumpling at others.

PLAYA A land-locked lake which dries periodically, leaving a crust of *evaporites* on the surface.

RAISED BEACH A beach now above the present sea level, formed at a time when the sea level was higher.

RED CLAY The deepest ocean deposit, formed from fine grained sediment carried in suspension or wind blown.

ROCK Any large, solid mass at the Earth's surface, composed of one or more *minerals*.

ROCK CYCLE The continuing transformation and movement of rocks on and in the Earth over time.

SABKHA A salt-encrusted coastal flat, flooded occasionally by the sea.

SALTATION The hop-like movement through air or water of particles which are lifted by the flow but are too heavy to stay in suspension.

SAND A sediment consisting of grains generally visible to the naked eye between 2mm and .063mm in size.

SCREE (or *talus*) The accumulated frost-shattered debris of rock fragments.

SEDIMENT Loose debris of *weathering* and *erosion*, transported to settle elsewhere.

SEDIMENTARY ROCK One of the three main rock types (with *igneous* and *metamorphic*), formed from sediment – debris of rocks or organic remains, or precipitating from solution.

SHALE A *mudrock* which breaks up in a flaky fashion

SILT A sediment consisting of grains not visible to the naked eye that are intermediate in size between those of *sand* and *clay*.

SINK-HOLE A feature of limestone scenery, where surface waters go underground.

SLUMP A type of *mass wasting* in which there is a sudden slip of sediment as a whole.

SOIL HORIZON A distinct layer seen in a *soil profile*, a region of particular biological and chemical activity.

SOIL PROFILE A vertical cross-section of soil from topsoil to bedrock.

SOLUTE Dissolved material.

STACK A rock pillar, free-standing in the sea; remnant of an eroded sea-cliff.

STALACTITE A hanging depositional structure, typically of precipitated calcium carbonate on cave roofs.

STALAGMITE A rising pillar-like depositional structure, typically of precipitated calcium carbonate and found on the floor of caves (see *dripstone*).

STRIATIONS Long scour marks along a boulder or rock-face made by other rocks dragged over it by a moving glacier or ice sheet.

SUBDUCTION ZONE A margin where one *tectonic plate* is dragged under another (the process of subduction).

TALUS See *scree*.

TECTONIC PLATE One of the 12 or so segments of the Earth's lithosphere which moves slowly across the planet's surface.

TILL A deposit left by a retreating or melting glacier or ice sheet. Typically unsorted, although *meltwater* may carry away the finer material.

TILLITE A rock formed from consolidated *till*.

TROPOSPHERE The lowest layer of the atmosphere, where major air currents are formed and *weather* occurs.

TSUNAMI A giant, ocean shock-wave following an earthquake or volcanic eruption.

TURBIDITE A rock formed from the deposited material of a *turbidity current*.

TURBIDITY CURRENT A cloudy mass of suspended sediment which moves like a current.

U-SHAPED VALLEY A wide, deep valley scoured by moving ice.

UNCONFORMITY A clear break in a sedimentary sequence, which marks a change in the history of that environment, e.g. when a period of erosion replaced a process of deposition.

UNIFORMITARIANISM The fundamental concept that geological processes operate consistently through time, so that knowledge of present processes can be used to explain the past.

V-SHAPED VALLEY The classic shape of a valley eroded by a river cutting down.

VARVE The layer of sediment laid down in a year, as applied to lake sediments which show clear annual cycles.

VENTIFACT A pebble with facets eroded by wind–sand abrasion.

VISCOSITY The tendency of material to resist flow ('stickiness').

WADI A valley carved by an intermittent stream or river in a usually arid environment.

WATER CYCLE The pathways and stores of water circulating on and within the surface of the Earth, including the atmosphere.

WEATHER The day-to-day conditions in the atmosphere affecting a particular place (see *climate*).

WEATHERING The physical fragmentation and/or chemical breakdown of rocks at or near the surface of the Earth.

FURTHER INFORMATION

The list below gives some books and other sources of information where you can continue to explore the topics covered in this book and find links to further sources of information.

BOOKS

Earth: The power of the planet, Iain Stewart & John Lynch, BBC Books, 2007. ISBN: 980-0-563-53914-8. Written to accompany a BBC TV series.

Earth surface processes, Phil Allen, Wiley Blackwell, 1997. ISBN-13: 978-0632035076. 416 pp. An introductory textbook focusing on the Earth's surface.

The map that changed the world: the tale of William Smith and the birth of a science, Simon Winchester, 2001. ISBN 0-670-88407-3. 338 pp. An account for the general reader of some of the early history of the science of geology.

Understanding Earth, 5th edn., John Grotzinger, Thomas H. Jordan, Frank Press and Raymond Siever, 2006. ISBN-13: 978-0716776963. 670 pp. An introductory text book.

WEBSITES

The rock cycle – a website aimed at UK science students (Key stage 3), showing how surface and deep Earth processes produce the rocks we stand on. http://www.geolsoc.org.uk/gsl/education/rockcycle

Earth in our hands – this is a series of 12 information notes, each addressing topics such as earthquakes, flooding, landslides, coastal erosion and highlighting the way geoscientists' work benefits society. Each leaflet suggests a range of further reading and useful websites. http://www.geolsoc.org.uk/gsl/education

A series of on-line public lectures on topics such as earthquakes, water resources and the geological record of climate change. http://www.geolsoc.org.uk/gsl/events/londonlectures

GENERAL WEBSITES

The following websites of the British Geological Survey and United States Geological Survey offer educational material and much other information concerning the Earth's surface.

http://www.bgs.ac.uk/

http://www.usgs.gov/

CLIMATE CHANGE

Climate change 2007: synthesis report – Summary for policymakers Intergovernmental Panel on Climate Change. This provides a summary of the IPCC's work set out for policymakers and other interested general readers. http://www.ipcc.ch/

Climate change controversies: a simple guide, The Royal Society. This addresses for non-experts some of the misleading arguments commonly made about climate change. http://royalsociety.org/page.asp?id=6229

The truth about climate change, DVD presented by David Attenborough, 2008, ASIN: B00179CXIK

PICTURE CREDITS

Front cover: © Adrian Warren/Lastrefuge.co.uk   Back cover: all images © Istockphoto

PREFACE, INTRODUCTION

pp4, 6 © AP/PA Photos; p.7 © Istockphoto; p.8 © Gary Hincks; p.9 top © DKimages; p.9 bottom © The Natural History Museum, London; p.10 © Nicole Heinzel; p.11 top Courtesy the British Geological Survey, © NERC; p.11 bottom © S. Jones; pp.12 top and bottom, 13 © Istockphoto; p.14 top, middle and bottom, and p.15 top (by permission of Gary Hincks) © Mike Eaton; p.15 bottom © DKimages; p.16 top from How The Earth Works, published by Dorling Kindersley; p.16 middle by permission of Dorling Kindersley Limited: How The Earth Works; p.16 bottom How The Earth Works published by Dorling Kindersley; p.17 top © Sally Alexander; p.17 bottom © Istockphoto; pp.18-19 © Istockphoto; pp.19, 20 © The Natural History Museum, London; p.21 © Istockphoto; p.22 top, bottom © The Natural History Museum, London; p.23 top, bottom right © Istockphoto; p.23 bottom left © The Natural History Museum, London.

WEATHERING

p.24 © Istockphoto; p.26 top © Tony Waltham Geophotos; p.26 bottom © The Natural History Museum, London; p.27 top © The Natural History Museum, London; p.27 bottom © William C. Jorgensen with permission from Genevieve B. Anderson at Santa Barbara City College; p.28 top © Istockphoto; p.28 bottom © The Natural History Museum, London; p.29 Courtesy the British Geological Survey, © NERC; p.30 © The Natural History Museum, London; p.31 © DKimages; p.31 © Jacques Jangoux/Science Photo Library; p.33 © Martin Bond / Photofusion Picture Library.

EROSION

p.34, 36, 37 top left © Istockphoto; p.37 top right © Tony Waltham Geophotos; p.37 bottom left © The Natural History Museum, London; p.38 © Istockphoto; p.39 top, bottom © DKimages; pp.40-41 © Istockphoto; p. 42 © Brian Romans; p.43 from How The Earth Works published by Dorling Kindersley; p. 44 top © Sipa Press/Rex Features; p.44 bottom left © Getty Images; p.44 bottom right © AP/PA Photos; p.45 top © Sipa Press/Rex Features; p.45 bottom © US Army/Science Photo Library; p.46 © Istockphoto; p.47 top by permission of Dorling Kindersley Limited: How The Earth Works; p.37 bottom © Mike Eaton; pp.48-49 © U.S. Department of Defense; p.49 © E.R.Degginger/Science Photo Library.

DEPOSITION

p.50 © NASA/Science Photo Library; p.51 © Gary Hincks; pp.52-53 © Sally Alexander; p.53 © Ecoscene; p.54 top © Tony Waltham Geophotos; p.54 bottom Courtesy the British Geological Survey © NERC; p.55 top © Istockphoto; p.55 bottom © Tony Waltham Geophotos; pp.56, 57 © Istockphoto; p.58 top © Tony Waltham Geophotos; p.58 bottom © Ocean Explorer/NOAA; p.59 © Istockphoto; p. 60 top © The Natural History Museum, London (redrawn from an original by Gary Hincks); p.60 bottom © Ingemar Renberg, Umeå University; p.61 top, middle © The Natural History Museum, London; p.61 bottom © Kurt Hollocher; pp.62-63 © Martin Bond/Science Photo Library; p.63 top © The Natural History Museum, London; p.64 Courtesy the British Geological Survey © NERC; p.65 © Gary Hincks.

FROM SEDIMENT TO ROCK

p.66 © Istockphoto; p.67 left © The Natural History Museum, London; p.67 right © Mike Eaton (photo © The Natural History Museum, London); p.68 © Istockphoto; p.69 top, bottom © The Natural History Museum, London; p.70 from Earth Atlas published by Dorling Kindersley; p.71 © DKimages; p. 72 top © DR Ken Macdonald/Science Photo Library; p.72 bottom © Mike Eaton; p.73 © The Natural History Museum, London; p.74 left, right © The Natural History Museum, London; p.75 © Gary Hincks; pp.76-77 © Tony Waltham Geophotos; pp.78-79 © Istockphoto; p.79 Courtesy the British Geological Survey © NERC; p.80 © DKimages; p.81 © Mike Eaton.

ROCKS AND THE LANDSCAPE

p.82 © Istockphoto; p.83 © Tony Waltham Geophotos; p.84 © Perks Willis Design; p.85 © Tony Waltham Geophotos; p.86 top © Zazu Arnold; p.86 bottom © Istockphoto; p.87 © Zazu Arnold; p.88 top © Dae Sasitorn/Lastrefuge.co.uk; p.88 bottom © Tony Waltham Geophotos; p.89 © Istockphoto; pp.90-91 © Istockphoto; p.91 top Courtesy the British Geological Survey © NERC; p.92 left © David George; pp.92-93 © World Pictures/Photoshot; p.93 bottom © David George.

STUDYING CLIMATE CHANGE

p.94 © Istockphoto; p.95 © Mike Eaton; p.96 Model by F. Mélin © E.C. Joint Research Centre; p.97 © D.A. Peel/Science Photo Library; p.98 © Mary Evans Picture Library; p.100 top © Istockphoto; p.100 bottom © DKimages; p.102 © Mike Eaton; p.105 © On the Rocks – The Geology of Britain, British Broadcasting Corporation.

Every effort has been made to contact and accurately credit all copyright holders. If we have been unsuccessful, we apologise and welcome correction for future editions and reprints.

INDEX